GINSENG DREAMS

GINSENG DREAMS

THE SECRET WORLD OF AMERICA'S MOST VALUABLE PLANT

KRISTIN JOHANNSEN

THE UNIVERSITY PRESS OF KENTUCKY

Publication of this volume was made possible in part by a grant
from the National Endowment for the Humanities.

Scholarly publisher for the Commonwealth,
serving Bellarmine University, Berea College, Centre
College of Kentucky, Eastern Kentucky University,
The Filson Historical Society, Georgetown College,
Kentucky Historical Society, Kentucky State University,
Morehead State University, Murray State University,
Northern Kentucky University, Transylvania University,
University of Kentucky, University of Louisville,
and Western Kentucky University.

Editorial and Sales Offices: The University Press of Kentucky
663 South Limestone Street, Lexington, Kentucky 40508-4008
www.kentuckypress.com

06 07 08 09 10 5 4 3 2 1

Photos and book index courtesy of Kevin Millham.

Illustration courtesy of Dan Dourson.

Library of Congress Cataloging-in-Publication Data

Johannsen, Kristin, 1957-
Ginseng dreams : the secret world of America's most valuable plant
/ Kristin Johannsen.
p. cm.
Includes bibliographical references and index.
ISBN-13: 978-0-8131-2384-4 (hardcover : alk. paper)
ISBN-10: 0-8131-2384-4 (hardcover : alk. paper)
1. American ginseng. I. Title.
SB295.G5J64 2006
633.8'8384—dc22
2005034193

This book is printed on acid-free recycled paper meeting
the requirements of the American National Standard
for Permanence in Paper for Printed Library Materials.

Manufactured in the United States of America.

AAUP Member of the Association of
American University Presses

CONTENTS

ACKNOWLEDGMENTS

I am tremendously grateful to all of these people, who were so generous with their time, knowledge, and insights: Gary and Beth Anderson, Kirk Baumann, Andrew Bentley, Dan Bond, Steve Bowling, Wendy Cass, James Chamberlain, Rock Cianciotti, Clare Comer, Jim Corbin, Pat Ford, Father Al Fritsch, Steve, Diane, and Seth Goodman, Joe Heil, Paul Hsu, Mary Hufford, Terry Jones, Father Jack Kieffer, Sherry Kilgore, Chris Kring, James McGraw, Laura Murphy, Scott Persons, Iris Riesen, Ann Rogers, Ginger Shelby, Sister Therese Tackett, Robert Watts, Skip Wissinger, Jo Wolf, and Syl Yunker. Wherever they speak in this book, their words have been quoted directly from recorded conversations. I have edited some of the conversations for clarity and brevity.

I greatly appreciate the help and hard-to-find information given to me by the Appalachian Ginseng Foundation, Appalachia-Science in the Public Interest, Berea College Archives, the Ginseng Board of Wisconsin, the Korean National Tourism Organization, the Lloyd Library of Cincinnati, Ohio, and the Marathon County Historical Society of Wausau, Wisconsin.

Special thanks go to my "literary critics," Karen Johannsen-Talsky and Bob Cabin, who read and critiqued earlier drafts.

And above all, I would like to thank my parents, Walter and Marilyn Johannsen, for their support, and my husband, Kevin Millham, for his encouragement, and all those dinners, and for giving me that calendar in the first place.

May they all be well and happy!

INTRODUCTION

SCENES FROM THE GINSENG WORLD

Ko Shing Street, Hong Kong. It's a blazing morning in early September, with temperature and humidity both in the 90s. "Take lots of water," our friend Iris warned us before we descended from her air-conditioned haven on the twenty-eighth floor. We'd spent the night before in her glass-walled guest room, and could hardly bear to sleep, the view was so enchanting, through the neon skyscrapers that parade down the steep slope to the harbor, out over the procession of freighters and junks and hydrofoils skimming the dim sheet of water, across to glittering Kowloon and beyond to the mountainous bulk of the Chinese mainland.

Seen from above, Hong Kong is gleaming and futuristic, a cool, high-tech, glass-and-steel machine. But at street level, it's another matter entirely, steamy and seething with humanity. Just as was true many dynasties ago in China, each neighborhood and every street has its own trade, and in Ko Shing Street, a stone's throw from the world's second-tallest building, it is traditional medicines. A weird, chokingly sweet smell pours from the open fronts of the shops. Dried lizards are mounded next to heaps of huge, shiny tree fungus, chrysanthemum blossoms pressed into ruffled cakes, bundles of powdery sticks, and rows of earthenware jars crosshatched with Chinese characters.

Previous page: On Ko Shing Street in downtown Hong Kong, employees of a traditional Chinese medicine company sort and grade heaps of ginseng root.

But mostly, there's ginseng.

Barrels of it, sacks of it, bales of it stitched up in canvas. The other name for Ko Shing Street is Ginseng Street, and more of it is pouring in even as we gawk. By 9 A.M., the street's two lanes have become a parking lot for delivery trucks, and sweating laborers with towels wrapped around their heads are wrestling giant bales into the shops, or barging down the sidewalk with handcarts and shoving guys in suits out of the way. I nearly get bashed in the face with a sack as I cozy up hastily against a wall.

Many of the businesses on this street are ginseng wholesalers, each shop like a long, narrow tunnel stacked floor to ceiling with huge cardboard cartons—a king's ransom worth of the roots. Far at the back, there is a single desk, with a figure hunching intent over papers and calculator. In one office, I see three men in suits lean forward over a glass display counter, pointing and arguing with the proprietor behind it.

The retail shops are more colorful, stacked deep in gaudy red and gold cardboard boxes of ginseng roots, ginseng slices, ginseng powder, ginseng tonics. Some frilly, ancient roots are individually displayed in packages that look like picture frames. And many of them proudly, conspicuously bear the label "American ginseng"—decorated with the Stars and Stripes.

This Hong Kong street is the world's main market for American ginseng. In 2000, more than three hundred tons of ginseng from the United States changed hands here, from Midwestern farms and Appalachian hollows.

It's a massive operation. In one of the shops, five men and women are seated at a long table buried under ginseng roots. I watch as they deftly break off the root-hairs, snap the ginseng into inch-long pieces, and then toss it into plastic baskets, dividing it by grade according to a system that is a tightly guarded secret. No one even sees me.

For over a billion Chinese—one in every six of the Earth's people—ginseng is the supreme medicine, one which strengthens every system of the human body, preventing disease and promoting longevity. The roots that arrive in this hot, noisy little street in Hong Kong will be processed, packaged, and shipped off to stores and pharmacies all across China, and to the Chinatowns of major cities around the world.

In the Chinese system of medicine, ginseng is held to stimulate the flow of *chi,* or life energy, throughout the organism. In combination with other herbs, it is used to treat everything from heart palpitations to deafness to life-threatening fevers. But in keeping with the overall philosophy of traditional Chinese medicine, which seeks to balance disharmonies in the body before they become full-blown diseases, the principal use of ginseng is as a tonic to remedy deficiencies in energy flow. Across Asia, hundreds of millions of people consume ginseng every day of their lives.

Though you'll find ginseng in just about any drugstore in North America, and natural foods enthusiasts are becoming increasingly aware of it, the demand for ginseng is still overwhelmingly Chinese. The Chinese have treasured ginseng for millennia—since before historical records began—and even the surge of modernization in their country in the last fifty years has not hurt demand. In fact, with more money in their pockets, even many poorer Chinese can now afford it. Demand seems insatiable.

It disorients me to raise my eyes from this almost medieval scene in Ko Shing Street—the stacks of bird's nests, the dead lizards, the barrels of ginseng roots—up to the glass-and-steel towers of investment banks and corporate headquarters soaring overhead. Something is making me very dizzy. It might be the vapors of a dozen potent medicines, or it could just be the heat. I decide that my friend Iris was right—it's just too hot to run around outside—and I scurry back to the ice-cold, indoor world of twenty-first-century Hong Kong.

A world away, and twenty miles from my home in Kentucky, I stand holding a plastic bag full of seeds. They don't look like much. They're a damp dark brown, the size of lentils, but lumpy and irregular in shape, bulging as though about to launch into life. In fact, some have already split open their seed coat, and a line of white, curved thin and sharp as a fingernail paring, shows where the first shoot is set to emerge.

"I brought plenty for everyone," Sister Therese Tackett tells the folks crowding around her at the ginseng growing workshop. "And if you want extra, it's ten dollars for a quarter-pound. Just remember, you have to keep those seeds in the fridge till you're ready to plant." Checkbooks are emerging, billfolds unfurling, and expectancy charges the air.

After listening to the presentation, I can't help running the numbers in my head. A quarter pound, that's twelve hundred seeds. If I planted them, about 85 percent of them would eventually come up; after growing and dying back for three years, the plants would start forming bright red berries, each holding two or three seeds. If I stored those new seeds in just the right conditions and planted them, I could start another field, and another. If I had just planted the contents of this bag ten years ago and walked away, with a certain amount of luck I'd now be harvesting a crop worth $500 a pound. An acre of it could bring me a cool $70,000.

No wonder everyone is mobbing Sister Therese. There's a farmer from Harlan there with his six-year-old grandson, a woman from Frankfort named Fatima, a pair of hippie herbalists in long skirts, three generations of a family who live just down the road, here in Rockcastle County. Some estimate that up to ten thousand people in the eastern United States have already planted seeds like these, and are now quietly, almost furtively, biding their time till the harvest.

American ginseng is the most valuable wild plant on our continent—and one of the most untamable. Botanically, ginseng is part of the *Araliaceae* family, seven hundred species which also include, oddly enough, celery, parsnips, and carrots. There are between five and eleven species of ginseng, depending on which taxonomist you listen to, but only two species, closely related, have important medicinal uses. *Panax ginseng*, which once grew wild from Siberia into China and the Korean peninsula, is known as Asian ginseng, while its cousin, *Panax quinquefolius* (or sometimes *quinquefolium*), which flourished in the wild from Quebec to Georgia and west to the Mississippi River, is known as American ginseng. The two plants look quite similar, and have similar active ingredients, but in the Chinese system of medicine they are held to have quite distinct (and complementary) properties.

American ginseng is a very odd plant. It has one of the longest germination periods of any known. When its small crimson berries fall to the ground in autumn, the seed inside remains dormant through the winter, the next spring, an entire growing season, and then another winter. It finally sprouts a year and a half after it fell. This was what had persuaded centuries of farmers on two continents that ginseng was impossible to cultivate.

Once awakened, the plant grows with excruciating slowness. The first year, it squeezes out a grand total of three leaves, only to die back again in fall, leaving just a small root with a bud on top that will send up a new plant the next spring. In the second year, it manages to put out two branches with leaflets. In the third growing season, it may finally flower, a cluster of small white blossoms that give way to berries the size of large peas. A fully mature plant is around twenty inches tall, and develops a maximum of five branches (or prongs) with a cluster of leaves. "Five-prongers" are the stuff of legend and barroom bets.

Far more important is what is taking place beneath the ground. The root continues to develop each year, its color darkening and its shape thickening and branching as it forces its way through the soil of the forest floor. The finest ginseng roots, according to the Chinese, are those that take on the form of a human body, with a torso, arms, and legs, and eerie threadlike appendages branching further. Every year, the plant's stem breaks off in the fall, leaving an eye-shaped scar on the side of the root. This makes it possible to determine the age of a root precisely. In the past, it was not unusual to find plants with the marks of twenty years' growth in the wild; some have been discovered nearly a century old.

Ginseng is very fussy about where it will grow. There must be at least forty days of below-freezing temperatures in the winter, or the seeds will not sprout. It needs soil that is at once well-watered and well-drained—the climate must be cool and temperate, with plenty of rain, but the water must not stagnate around its roots. For this reason, a steep hillside is best. It needs protection from the direct rays of the sun by an unbroken canopy of forest. It needs soil that will keep the roots growing slowly as they force their way down. And most of all, it needs a particular combination of nutrients found only in the decaying leaves of certain hardwood trees that drop their leaves early in the fall, such as maples and poplars.

In all of human history, only two environments in the world have offered these ideal conditions for ginseng, and one of them has vanished. The first was the forests of northern China, which once stretched from the Korean Peninsula across China and into Siberia. Most of this forest was cut down over a millennium ago, and even if the land were miraculously reforested today, the soil would take many centuries to reproduce. The other area still remains: the forests of the Appalachian Mountain chain in eastern North America.

American ginseng has a long and venerable history. It was America's first export to China—before our nation even existed. Fortunes were made, and lost, on ginseng. John Jacob Astor, the founder of one of America's great financial dynasties, got his start in the ginseng trade. Daniel Boone, the trailblazer of America's westward expansion, lost a fortune on it. Generations of families in America's eastern mountains, from New York to Georgia, relied on wild ginseng as a source of cash income, making it ever scarcer in the wild. Today, the plant's forest habitat is under siege, threatened by mining and logging and urban sprawl.

Over a century of attempts at cultivating ginseng, from the 1870s onward, have produced a sizable income for some, and nothing but heartbreak for others. Ginseng has been raised intensively in settings as far-flung as Australia, British Columbia, the mountains of Korea, and central Wisconsin, but farming produces roots very different from the wild variety, with only a fraction of the value. In recent years, a number of individual farmers have been quietly working out techniques that produce roots that are indistinguishable from wild ginseng—and which fetch the same breathtaking prices.

Today, ginseng is the most valuable forest product in the world. Top-quality wild roots sell for up to $1,000 a pound in that hot, noisy street in Hong Kong, and a wealthy ginseng lover might pay up to $10,000 for a single perfect, man-shaped root. Ginseng from America's forests is a major export, amounting to more than $24 million per year. China's surging economic development is raising hopes of an ever-expanding market for ginseng.

With her plastic bags of seeds, Sister Therese Tackett is just one of many people who feel that ginseng could be the salvation of the forest—and of Appalachian farmers. Because the costliest ginseng grows only in undisturbed forest, with its canopy and leaf litter intact, the astronomical value of the roots provides a powerful incentive to preserve the vanishing woodlands of the eastern mountains. It also offers the promise of a healthy income to rural families in a region long plagued by unemployment and poverty. It sounds like a gold mine, but even Sister Therese will tell you: There are a few problems that need solving first.

There's a Chinese legend that says ginseng is the child of lightning.

Up in heaven, water and fire fight an eternal struggle, two opposing elemental forces trying to conquer the universe, and they pour down rain, snow, and hail on the world—and blast it with lightning.

And if that lightning happens to strike a spring of water, the water will disappear into the earth, and in its place a ginseng plant will grow, blending yin and yang, water and fire, darkness and light, embodying the life force that moves the universe.

Many people say that ginseng has magical powers. In Korea, there are tales of a ginseng root that becomes a little boy, or grants wishes, or talks, or turns itself into another plant. In Siberia, the tribes that hunted it believed ginseng would show itself only to a pure and moral man. They said prayers before setting off. In the hollows of Appalachia, folks still swear that the plant hides itself under the ground when it knows it's in danger, and comes up again only several years later.

I can say for certain that it casts a spell. It surely has on me.

The first time I tasted ginseng was on my second day in Korea. After finishing graduate school, I packed up my master's degree and got a job teaching at a private language institute in Seoul. It was 1987, during the run-up to the Olympics, and English teachers could just about write their own paychecks. The day after I arrived, my new (American) boss invited me for lunch, as he did to welcome all the new teachers, and he took me to the same restaurant where he always took them. It was across the alley from the school, in a district where new glass-walled office buildings seemed to launch themselves skyward in a week's time. The restaurant had a shining white tile front, and served just one dish—*samgyet'ang,* ginseng chicken. They brought it out in a heated bowl made of black stone.

"I thought ginseng was a kind of medicine," I said, cautiously picking into the whole stewed hen with the blunt wooden tips of my chopsticks.

"Actually, it is," my boss told me. He'd come to Korea as a Peace Corps volunteer and never got around to leaving. "But Koreans think it's good for everything. So you can never have too much of it."

The dish was tasty, if surprisingly bland for Korean cuisine. The hen swam in a pale, mild broth, and it was stuffed with mushy white rice, whole date-like fruits, and a long, yellowish root that resembled a skinny carrot boiled too long. I fished the ginseng out with my chopsticks and took a hesitant nibble. It was soft, bitter, and tasted older than dirt.

Over the next three years in Korea, I ran into ginseng everywhere I went. Ginseng grew in low, shaded sheds along the highway I took every morning to get to my second job at a suburban university. There were stubby brown glass bottles of ginseng tonic by the cash registers of every pharmacy, stored in little heated cabinets that kept the brew at a toasty temperature. There were vast mounds of raw ginseng root on tables and on tarps on the ground at the sprawling outdoor market in Jongno O-ga, just around the corner from one of Seoul's big, modern department stores. When I took an unnerving bus tour through the Demilitarized Zone at the North Korean border, where the Korean War was fought to a permanent stand-off, there were tell-tale rows of the long sheds where it grows. During our last week in Korea, one of my new husband's best friends gave us a very special present to take home: a tiny white porcelain jar containing pure extract of ginseng.

I went to visit my sister, back in Wisconsin. She had moved from our hometown of Milwaukee up to Stevens Point, dead center of the state. After a couple years studying forestry at the state university there, she switched her major to business—giving her slightly more hope of paying off her student loans before middle age—but she was treating herself to an elective course in horticulture. She told me they took great field trips to local farms—a potato farm, a cranberry bog, a place where they raised gladiolus bulbs (all in bloom!), a ginseng farm—

"*Ginseng?*"

"Yeah. I guess they sell it to Chinese people. Anyway, it's a really cool class."

I filed that away, too.

Fast-forward a decade. My husband and I had just moved to a small town on the edge of the Kentucky mountains, where I ended up writing articles about local issues—tourism problems, the Army's chemical weapons depot ten miles up the road where nerve gas was leaking from obsolete rockets. For a birthday present, Kevin brought me a calendar with scenic photos of Appalachia and information on environmental groups in the area.

One of them was called the Appalachian Ginseng Foundation. I was puzzled enough to pick up the phone.

That was the beginning of a four-year odyssey that's taken me from muddy Appalachian hillsides to a gleaming cancer research lab, from skyscraper streets in Hong Kong to the ranger station of a national park, chasing the truth about ginseng.

And all the way, I heard the most remarkable stories—tales of hope, and loss, and dreams.

ONE

MAN-ROOT

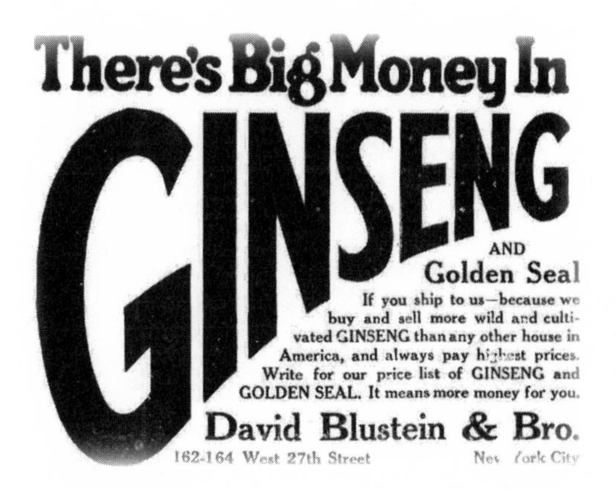
There's Big Money In
GINSENG
AND
Golden Seal
If you ship to us—because we buy and sell more wild and cultivated GINSENG than any other house in America, and always pay highest prices. Write for our price list of GINSENG and GOLDEN SEAL. It means more money for you.
David Blustein & Bro.
162-164 West 27th Street
New York City

In November of 1912, a peculiar little magazine made its first, unheralded appearance. It was the brainchild of an elderly man named Penn Kirk, the owner of a carpet-weaving business in Arrowsmith, Illinois. He served as its editor, publisher, sole reporter, desultory proofreader, subscription agent, advertising salesman, and chief promoter. He called it *The Ginseng Journal.*

A handful of farmers scattered around the eastern United States had begun experimenting with cultivating the valuable plant as much as thirty years earlier, and by now they figured they were getting it down to a science. Through endless trial and error, they had learned that the seeds needed a year and a half and the sharp cold of winter to germinate, and that the plant flourished best under the canopy of the forest but would grow, grudgingly, under artificial shade. The idea had traveled by word of mouth, until rumors of easy riches from ginseng growing spread from the forests of Maine to the mountains of California. This was when Penn Kirk decided to launch his magazine.

It wasn't a very impressive debut. The cover of the *Ginseng Journal,* volume 1, number 1, is adorned only with the subscription price ("Six months 25 cents One year 50 cents") and a list

---

Previous page: This advertisement was published during the first great boom years of American ginseng, in the August 1916 issue of the *Ginseng Journal and Goldenseal Bulletin.* Courtesy of Lloyd Library and Museum.

of advertising rates by the word, inch, or page. Its first two pages contain (crammed between ads for a "bottle clothes sprinkler," asthma cures, and a magazine called *Modern Electrics*) recipes for invisible ink and sneezing powder. It isn't until page five that we reach the heart of the matter. There we read how an Ohio farmer struck it rich: "George Wilson, of the East Pike, Wednesday received a check for $1,102 for the balance of his last crop of the medicinal root just marketed by him in China. Mr. Wilson says he spent only $25 in cultivating the roots sold for this sum. He secured 4,300 per cent on his investment." With a 1912 dollar the equivalent of $19 today, the fortunate Mr. Wilson had just raked in over $20,000.

Of the sixteen pages in the magazine's first issue, only two and a half concern ginseng. The rest are a three-page installment of a serial, floating in an ocean of ads for get-rich-quick schemes (compiling mailing lists, agents to sell a "household necessity," and that perennial favorite, stuffing envelopes). Kirk's "Talks on Ginseng Raising" are heavy on enthusiasm, though skimpy on facts: "Ginseng," he declares, "is of slow growth, and therefore all the more valuable. A tree is also of slow growth, but timber is valuable anywhere you use it. When you consider how quickly you can begin to sell seeds from your seng, and how rapidly the price advances for the plants as they get older there is no time that they depreciate in value. Bankers will loan money on a growing crop of ginseng, and your credit is strengthened with them as soon as they know you are carefully growing it." Logic and rhetoric were not his strong suits.

But the magazine took off. The second issue is printed on lovely pink paper, and its cover bears a photo of an enormous mound of harvested ginseng root and the slogan, "Devoted to the Interests of Ginseng and Golden Seal Growers" (goldenseal is a less valuable herb still cultivated alongside ginseng). In his Ginseng Talk, Kirk answers a couple of readers' letters asking about preparing the soil ("First clean it well and next spade it well. Third drain it well.") and the issue of starting from seed vs. buying plants ("Let me say that if you have ready money never think about spending the time fooling with seeds."). The magazine discusses the process of stratifying seeds, before moving on to a page on goldenseal, an article on the history of the Christmas tree, two page-long poems, a short story, and the inevitable ads for "How to go on the Stage," a home-study taxidermy course, starting one's own tooth-powder manufacturing business, and something called the "Endless Dime Scheme" (for which, suspiciously, you are to send ten cents).

Penn Kirk lived in a vastly different America, one populated by fewer than 100 million people, a land in which many had never been farther from home than they could travel in a horse and wagon. Though automobiles were becoming more common—more than half a million of them were already on the road—there were very few places to drive them. Not a single mile of paved road existed outside the major cities, and the Lincoln Highway, the first all-weather road across the United States, was still nothing more than a very hypothetical line across the map. Covering a distance of ten miles could be a day's work on the rutted, impassible roads, where auto and horse-cart alike could bury themselves to the axles in mud after a hard rain.

And yet, in the forests of Minnesota, in the hills and hollows of the Tennessee mountains, in backyard gardens in quiet Pennsylvania towns, farmers were growing a plant for export to a dimly known country halfway around the world—imperial China. How this came about is an implausible tale that depends, ultimately, on the scientific curiosity of two men in the early 1700s. They were both French Jesuit priests, and they never met each other.

The Society of Jesus, the official name of the Jesuit order, was founded in 1534 by St. Ignatius of Loyola to bring new life to the Catholic Church following the Protestant Reformation. These highly educated priests had as their mission education and scholarship, and often focused on converting the elite of the countries they visited to Christianity. One Jesuit strategy was to use Western science to perform services for the local rulers. It was with this aim that a group of Jesuits was dispatched to

China to compile the country's first accurate atlas for Emperor Kang Hsi. Among them was the French mathematician Father Pierre Jartoux. Along with his surveying work, Jartoux compiled reports on things Chinese for the use of his order. In 1711, he submitted a report to the Jesuits' Procurator General entitled "The Description of a Tartarian Plant, called Gin-seng, with an Account of its Virtues."

These virtues, he believed, were many. "Nobody can imagine that the Chinese and Tartars would set so high a value upon this root if it did not constantly produce a good effect. . . . It is certain that it subtilizes, increases the motion of, and warms the blood, that it helps the digestion and invigorates in a very sensible manner." He tried it himself as a remedy for exhaustion, and found that "an hour later I was not the least sensible of any weariness."

Jartoux, a meticulous scientist, went on to record in careful detail how the Chinese prepared the plant by slicing and boiling it. He described the soil and climate conditions that favor its growth, and the harsh life faced by wild ginseng hunters in Manchuria: "They carry with them neither tents nor beds, everyone being sufficiently loaded with his provision, which is only millet parched in an oven, upon which he must subsist all the time of his journey. So that they are constrained to sleep under trees, having only their branches and barks, if they can find them, for their covering."

But most importantly, Jartoux gave a finely observed description of the plant's appearance and growth habits, and accompanied it with a precise scientific drawing along with measurements. Though Europeans had been aware of ginseng since the time of Marco Polo's first descriptions of China, no one knew the plant's precise nature, what the Chinese used it for, or where it grew. Father Jartoux's report was so invaluable that it was later translated and published in 1713 by the Royal Society of London, the preeminent scientific organization of the day.

But that was not where the second Jesuit learned of it. Father Joseph Francois Lafitau was a missionary to the Iroquois of Canada, likely a far less prestigious assignment than that of cartographer to the Chinese Emperor. He arrived in Quebec in 1715 to work at the Jesuit mission of St. Francis Xavier. Every year, his order would send out to its missionaries a collection of letters from their Jesuit brothers at work in other countries, containing practical and scientific information that could prove useful in their labors.

Father Lafitau was particularly interested in medicine. "In order to announce the truths of our religion to uncivilized nations," he wrote in a memoir, "and make them taste of a morality very opposed to the corruption of their hearts, we first have to win them over, and to insinuate ourselves into their spirits by becoming necessary to them. Several of our missionaries have succeeded in different places because they have a certain talent for medicine. I know that while working to cure the ills of the body they have been lucky enough in some cases to open the eyes of the soul." This, he decided, would be an excellent approach to use with the Iroquois, and he applied himself diligently to the study of medicine. When Father Jartoux's report on ginseng arrived in his bundle of missionary letters, he decided it sounded very promising indeed. A plant that powerful could help win any number of souls for Jesus.

Studying Jartoux's description of the soil and climatic conditions favorable to ginseng, he realized that his region of Canada, between Montreal and Ottawa, fit the bill perfectly. But the native people of the area didn't recognize the plant from the picture he showed them, and he spent three months combing the woods for it with no success. Finally, in late summer, a cluster of crimson berries caught his eye in the woods near a house that he was building. He showed the plant to a woman at the mission, and she recognized it as one of their minor remedies.

Immediately, Lafitau sent some samples of ginseng off to a plant expert in Quebec City, who agreed that it must be a form of Jartoux's plant. He then dispatched a report and samples to Paris. But saving souls with ginseng soon became a secondary consideration, as the French quickly grasped the commercial importance of the plant. Not long afterward, a trial shipment of ginseng roots was sent from Montreal to China. When the ship finally returned, after long months, the news was good. The Chinese had received the new form of ginseng enthusiastically, and deduced that its medicinal properties were somewhat different from those of the Asian root. Mission Indians were sent out in the woods to harvest the wild plant for shipment halfway around the world.

Father Lafitau returned to France in 1718 to attend to business, but he never went back to his mission in Quebec. Though he wanted to return to Canada, his superiors instead gave him the position of Procurator for all the missions in Canada. He later wrote several books about the Canadian Indians. It is not recorded whether ginseng ever brought any of their souls to the True Faith.

But why didn't China simply grow its own ginseng? In a word, it couldn't—and still can't. Ginseng is native to cool, shady forests, and requires deep soil made of centuries of decomposed leaf litter. Without the nutrients from the leaves, ginseng fails to flourish, or simply won't come up at all. More than a millennium ago, the pressure of China's burgeoning population brought the destruction of its last remaining temperate forest. Wild ginseng's habitat in China has been eradicated, and cannot be rebuilt, leaving them with no choice but to obtain it from Korea, Manchuria—or North America.

The ginseng trade in Canada quickly boomed. By 1720, ginseng was being exported to China on consignment by the Company of the Indies, a French trading company. Native Americans and white trappers combed the forests for it. Fur traders turned to it as a sideline, dealing in pelts in the winter and spring, and ginseng in the summer and fall—a pattern that continues in Appalachia to this day.

Gradually, the hunt for ginseng spread southward throughout the plant's natural range, and on into what is now the United States. In colonial times, the Catskill and Allegheny Mountains were prime hunting grounds, and Albany, New York, was an important trading center.

Daniel Boone was one notable ginseng hunter. While exploring and hunting west of the mountains in the newfound frontier of Kentucky, he gathered ginseng roots, along with the furs that were the mainstay of his livelihood. But as with just about every other facet of his life, his ginseng hunting was dogged with almost relentless bad luck. By one account, Boone had spent a year amassing a huge stockpile of roots at a time when the price was quite favorable, and loaded it on a keel boat to take it to market on the Ohio River. While attempting the cross a strong current, the boat overturned, soaking its cargo. He had to wait for help to come from three miles away, by which time the roots were mostly ruined. He attempted to dry the ginseng quickly in the sun—further damaging it—and ended up selling the roots for a pittance.

Some of these early ginseng tales, however, have to be taken with a grain of salt. Most writers quote each other and claim that Boone's boat carried "twelve tons" of ginseng—a staggering amount, enough to fill 3,000 barrels and outweighing all the wild ginseng collected by 8,977 diggers in the entire state of Kentucky in 2003. As folklorist Warren Roberts has pointed out, Boone was more likely transporting twelve tuns—an old name for a barrel.

But whatever the size of Boone's cargo, it's indisputable that large amounts of wild ginseng were shipped from North America. It was the continent's first export to Asia, and after the independence of the United States, a mainstay of the new nation's economy. In 1784 a ship called the *Empress of China* set sail for Canton, completely loaded with ginseng root. John Jacob Astor, the fur and land magnate, got his start in business in 1786 by buying every scrap of ginseng root he could find and chartering a ship to take it to China. From this venture he received $55,000 in silver coin—equivalent to $1,140,000 in our day, and the stake that founded a trading empire that stretched from New York to Oregon.

Ginseng expert Scott Persons believes that the plant's key role in America's westward expansion has yet to be appreciated.

"As people began to settle the land," he says, "from the Eastern Seaboard all the way out west, to Minnesota and Missouri and Iowa, in many areas ginseng was a very important part of that, because it provided immediate cash income to the settlers until they could get a crop going. They sold it in order to buy the shovel, to buy the flour."

In the Big Woods region of Minnesota, Persons tells me, there was a punishing economic depression in the 1860s. Crops couldn't be sold, and settlers were desperate for income. Just as the homesteaders were beginning to give up and move to Iowa in search of better times, a group of ginseng buyers from Virginia arrived. "People were going across the woodlands almost at arm's length digging ginseng. And it saved that section of Minnesota." For three years, out of the 350,000 pounds of American ginseng exported to China annually, about 250,000 came from Minnesota. "They had ginseng festivals, ginseng balls, somebody wrote a ginseng polka and they danced it. It was like the rain came after a drought. It saved these people." It also led to some of the earliest ginseng-related laws, when the Minnesota legislature passed measures to control the digging, but these proved to be too late. Within four years, the green gold rush was over, moved to other locations where the plant was still plentiful.

In his 1908 treatise *Ginseng and Other Medicinal Plants,* A. R. Harding of Ohio records the desperate poverty that drove some of the early diggers. He recalls, a number of years earlier, his having seen "one party of campers where the women . . . had simply cut holes through calico for dresses, slipping same over the head and tied around the waist—not a needle or stitch of thread had been used in making those garments." While some diggers traveled on horseback, "Others travel by foot carrying a bag to put ginseng in over one shoulder and over the other a bag in which they have a piece of bacon and a few pounds of flour."

Throughout the nineteenth century, American ginseng remained a lucrative business. In 1841, clipper ships carried over 640,000 pounds of dry ginseng to Asia. It is estimated that more than 60 million pounds of fresh ginseng roots were dug between 1783 and 1900. But ginseng was gradually becoming harder to find. As America's population grew and white settlers pushed westward, more and more forest was cleared for farming, and more and more hunters were out digging the roots, from Maine to the Mississippi and beyond. By the late 1800s, it was widely acknowledged that wild ginseng was on the verge of disappearing. "It is becoming very scarce, and, unless a method of cultivation becomes practical, bids fair to be exterminated," stated the 1909 edition of *King's American Dispensatory,* a pharmacy reference manual.

And so it was that a scattering of growers began trying to tame "nature's wildest child," attempting to persuade it to grow in neat rows under wooden screens. Penn Kirk, of Arrowsmith, Illinois, was hardly the first, or the most successful, but he was probably the most entertaining.

The back cover of every issue of the *Ginseng Journal* features a blurry photograph of "The KIRK Ginseng gardens at Arrowsmith, Illinois"—but what is clear is that it's an enormous spread. Wooden poles and lath screens stretch off into the distance, and underneath them rows of plants flourish on curved, raised beds. "Ginseng raising is like the raising of any other crop," Kirk proclaims. "It is the most economical of all crops, taking so little room and work. A well tended town lot of it would buy a farm. I sold $526.00 worth of it from a little space of 30x105 feet last year. Had it been left in the ground one year longer it would have doubled in weight. We learn by experience. . . . One man received $204.00 for one crop on a space sixteen feet square. He tended it well. It could have grown in the shade on the north side of his house. One acre thus tended would have yielded $33,000. See?"

We learn in the third issue that the *Ginseng Journal*'s endless pages of get-rich-quick ads ("Mink raising—fifty dollars from one mink yearly" and a book on "How to Become BEAUTIFUL") are actually placed by an agency, but Kirk's entrepreneurial spirit fits right in. "It will grow like weeds," he proclaims to his readers, "and we all grow some weeds. Just as well raise seng."

In another issue he trumpets, "If you want to wear diamonds, [and] eat pure buckwheat cakes with fresh sausage gravy over them for breakfast, just take the Journal for a single year and absorb the experience of every ginseng and seal raiser that you will read in it. Don't bother yourself about the stocks and bonds. But just sit in the shade, and pull weeds and wear the diamonds for a novelty. You can give them away for a present when you tire of them."

Within five months of its first appearance, Penn's magazine was thriving. "The Ginseng Journal now circulates in every state in the Union, Canada, Cuba, the Canal Zone, Prince Edward Island and Japan. How much better do you want it?" he boasts. Similar magazines had preceded it, including Carlos B. Paseador's *The Ginseng Garden,* which appeared from 1903 to 1906. Kirk's arch-rival, a ginseng publication called *Special Crops,* was published by Charles Goodspeed of Skaneateles, New York, for decades, from 1902 to 1936. But Kirk's *Journal* grew extremely popular for the variety of information it carried—and the testy opinions of its editor.

The March 1913 issue saw the first installment of a memoir in serial form called "The Old Log Cabin," written by the editor himself. It begins, "It was just 51 years ago when I made my first trip with father to the woods to hunt wild ginseng. I have been going to the woods about every year since then to find it. So you see there is more or less ginseng fever in my blood and make-up. Father was an ardent admirer of the wooded hills and hollows, and so am I." But soon, in Kirk's inimitable style, it careens off into ruminations on Abraham Lincoln and methods for teaching children to spell.

Mostly, "The Old Log Cabin" is a stream-of-consciousness outpouring, four solid pages of tiny type in most issues, about the Great Rebellion (otherwise known as the Civil War), dream interpretation, a proper country breakfast (fried mush), séances, his favorite scripture verses, or the unsatisfactoriness of the public schools, where children are allowed to read novels, he notes with horror. It follows no discernable order—certainly not chronological.

The serial continued in every issue of the magazine, and though it only occasionally meandered back to the topic of ginseng, readers could catch glimpses of the character of the old ginseng-hunter. Penn Kirk was born in a log cabin in the Shenandoah Valley of Virginia, one of fifteen siblings in a farming family. As a boy, he and his brother Thomas combed the woods for ginseng with the cannons of the Civil War booming in the background. Already the plant was getting harder to find: "Why we had this interest in woods life, and loved it enough to follow it so ardently there was no mortal could tell. But hither we would hie, spend long days out of sight of the rest of the world as it were, and listen only to the sounds that echoed in the tree tops from time to time as it re-echoed from the long whistling of a distant locomotive or steam thresher in a distant field. . . . Little by little we added to the mite of roots we were gathering, but at no time could we say we had an unusual good find unless it was to come to some spot where no hunter had ever been and roots were plenty."

As an adult, Kirk moved west to central Illinois, settling in Arrowsmith and opening a rug factory, and continuing to prowl the woods in search of increasingly scarce wild ginseng. He started a publication for the rug business, and when he began experimenting with ginseng cultivation, he naturally, and inevitably, shifted his literary outpourings to that industry.

Whatever he did, he grew obsessed with it, and fierce in his convictions. When one reader criticized his method of ginseng growing, he retorted: "Aside from its bountiful output is the one satisfaction of having been patient in the face of those who were inclined to make sport of my experiments. I have been unfortunate to be an inventor in my day and patented many articles for use in mechanics. Often my wife came to the shop very early in the morning to know why I did not come to bed. But she found me shivering in the cold experimenting. But she never questioned my sincerity like this one reader of the Journal."

A photo of the editor, which ran in later issues, shows a white-haired man buttoned tight in a dark wool suit. He is balding and clean-shaven, and wears tiny rimless spectacles. But his most prominent feature is his mouth, thin-lipped and so broad it seems to bisect his face. He looks like he's having trouble keeping it still long enough for the photographer to do his work.

At that time, little was known for certain about ginseng culture. Kirk's magazine became an invaluable forum for reports of experiments and speculation over how to improve growing methods. In an era when travel was difficult and telephones rare, the *Journal* brought monthly reports of pest problems, new shading methods, and soil improvement, from Canada to Kentucky to California. It carried the latest New York City market news to the most isolated farm in Appalachia. It also incited people in the most implausible places to attempt to grow ginseng. In one issue, Kirk assures a reader that the plant will do well in flowerpots at his home in New Jersey; he tells another reader confidently that there is no reason why ginseng shouldn't be profitably grown in (subtropical) Bermuda.

Within a few months, letters were flooding in from growers across North America, reporting their experiments, successes, and failures. The department called "From Our Correspondents" quickly grew to make up the bulk of the magazine. Readers discussed the best kind of mulch, whether to water the crop in dry weather, how to deal with diseases (a solution called pyrone was the recommended remedy), and most of all, how to produce the gnarled, twisted roots that Chinese buyers were paying the highest prices for. And to each letter, Kirk responded at great length—often three times as long as the original letter.

Some writers argued for transplanting wild roots into beds out in the forest. Others claimed that growing ginseng under artificial shade canopies produced better results, since the crop was easier to tend. Characteristically, Kirk expressed his view in no uncertain terms. At one point he received a notice from a company saying its "plants will be grown in the woods just as nature fixed it. The roots do not do near so well in ground artificially prepared." Retorted Kirk, "The last clause of the above is unstinted prejudice glibly expressed. It only comes from those who live handy to the woods and want to monopolize the ginseng business. . . . While it is more expensive to raise it artificially it is no reason it can't be profitably raised that way. You would just as well say you can raise a family better in the woods than in modern society. The inference is analogous and just the same." Left unsaid is the fact that unbroken forest was a scarce commodity in central Illinois, where he lived.

By this time, many growers had settled on a method of raising ginseng under screens built with wooden lath, to mimic the cool, dappled sun of the forest floor. The seeds were first stratified (left to rest) in containers of sand for a year, then planted in long beds that curved down gently from the center to a slight ditch, allowing water to drain off quickly. The plants were allowed to grow for four years or longer before they were dug, and their roots gently cleaned and dried. But there still remained much to debate about such things as pests, irrigation, and the most favorable soil types for the plant.

Meanwhile, Kirk was shipping off seeds to customers in Japan and South Africa, and enthusing over a scheme to produce ginseng chewing gum in North America, in order to develop a domestic market. A ginseng chewing gum factory was opened in Rochester, New York, enthusiastic reports were given of its booming business, and then the whole affair vanished from sight for a year. The proprietor, a Mr. Skeels, later reappeared, lamenting that he had been bilked out of $94,000, but declining to give the particulars.

And always there were the ads (autograph albums for 8 cents, bust developers, Texas Oil Land Opportunities, the Klamath Korn Kure for sore feet) and articles on "Outwitting a Ghost" and "Our Pet Hummingbird." Explained Kirk, "The outside matter spiced in is to give you a breathing spell and save the monotony of one subject alone."

For the Americans who grew it, ginseng was a cross-cultural encounter of the most baffling sort. Penn Kirk told his readers to sell leftover amounts and bits of low-grade rootlet fiber to the Chinese workers at their local laundry, then still a staple of the American townscape. He stated that these laundrymen will "use up your fibers like a rat would clean up the crumbs of cheese," and gave hints for dealing with the mysterious "Chink" (the lamentable but common epithet of the day): "He is rather cute [sly] in his bidding for it. You will never get him to offer what you want for it the first offer. He expects you to Jew down and you will have to ask more than you intend to take for it. Never take him at his first offer, for he will give you double what he first offers before he will let it go." This is followed by a "Recipe for Buns" sent in by Mrs. D. H. Terhune of Crescent, Oklahoma.

Most large growers either brought or shipped their harvest to New York City, where large-scale dealers, many of them Jewish family businesses, would buy the crops and then sell them to Chinese exporters. Kirk encouraged his readers to go to New York and negotiate with the dealers themselves, and a lengthy article in the December 1916 issue told them how to go about this. "THE TRIP TO MARKET TO NEW YORK CITY," it was titled. "What You Will Witness When You Go There Yourself."

From his home in Illinois, Kirk set off on the train with his daughter (a stenographer) and two large iron trunks crammed with ginseng. "We bought tickets to N.Y. via Cincinnati, Washington, and Baltimore," he reported, "and leaving our home town at one o'clock landed the next evening at nine o'clock in the heart of N.Y. So you see the trip was not as long and fatiguing as you might suppose it would be. . . . We got there Friday night and of course went to our rooms at once to get our breathing all right."

The next day, samples in hand, Kirk made the rounds of the various dealers, and there encountered many Chinese going from office to office trying to buy up huge quantities of the root at the lowest price possible. He met one who told a sob story of being robbed of $1,000 worth of precious old ginseng, in hopes of getting a better price from sympathetic listeners. Kirk ended up hiding his trunks so that the Chinese would not think he was a gullible farmer fresh from the countryside, and he told a highly implausible tale of two Chinese trailing him all around the city, from a dealer's office to a museum ("we visited the Egyptian Obelisk and noted the hierogliphics which we could no better read than the intentions of a Chinaman") to a streetcar. More likely, the Chinese truly did "all look alike" to him.

"They are as shrewd as yankees or Jews," he warned his readers. "They know how to drive good bargains if there is not another thing in the world they can do." Yet they won his grudging admiration: "The Chinks who are always on the lookout for snaps are to be commended for their shrewdness, and use their opportunities at all times."

Because politics had a direct (and often negative) effect on the price of ginseng, growers kept a careful eye on events halfway around the world. The *Journal* carried occasional analysis of developments in Asia that escaped the notice of the vast majority of Americans: "Political conditions in China are in a very turbulent state," it reported in June 1916. "Yuan Shi-Kai, who attempted to set up a republican form of government in China, later decided that the Chinese would be better off under a monarchical form of government and attempted to overthrow the republic and set himself up as Emperor. When this did not meet with the approval of the Chinese, he offered to change back to the republican form of government and call himself President. . . . should conditions in China change for the better and a permanent form of government, satisfactory to the people, be established, it is probable that the ginseng market will improve greatly."

And at the very foundation of the whole ginseng-growing enterprise was the eternal—and baffling—question: Why do the Chinese want this root so badly? And how long will the demand continue? Farmers discussed this endlessly in the *Journal*'s correspondence columns, seldom coming to any insight or conclusions. Wrote P. W. Krier from Wisconsin, "The 'Chink' has a bug house notion about the value of different roots that may be a thousand years old, and we are not going to change it." Kirk himself reckoned, "There are 450 million people in the Chinese Empire. If each one only used five cents worth of ginseng in a year it would require twenty-two and a half million dollars worth of ginseng to supply them. It would require two and one fourth millions pounds of ginseng to supply them if they paid ten dollars per pound for it, and over five hundred and ten acres to supply the product each year. They will buy all the ginseng we can raise for many years."

In an essay in April 1915, Kirk pondered the meaning of ginseng to the Chinese. He marveled at the fact that, of all the people in the world, only one nation holds the plant in such high esteem, and is willing to pay such enormous sums of money for it. Though he referred to the people as Chinks, and professed himself baffled by their ideas, still he admired them, and showed a certain respect for the beliefs of other cultures. "I speak of them as a nation, truly great in their preservation and oneness of sentiment. What there is among them of discontent they have only borrowed of those like ourselves. Century after century they have clung to their ginseng as no others have done. Why it is thus, seems to be the eighth wonder of the world. No other people seem to be like them. . . .

"The Israelite won't eat our hogs because he thinks them 'pizen' and only uses them for axle grease, but we use them in a thousand ways. We don't call him superstitious, but think he is about half right. It is certainly a freakish little concern this ginseng. Taking fully $20,000,000 to finance it, yet one half our 90,000,000 people do not know where it is and what it is for."

Which is Kirk's long-winded way of saying: To each his own.

Speakers of different dialects of Chinese have different names for ginseng, but they always write it with the same two characters, which mean Man-Root. For millennia the Chinese have been fascinated by the eerie way in which the root resembles a headless human body, with a slender neck and a thick torso branching into arms and legs, sometimes even fingers and toes. If like cures like, then such a root would be a peerless medicine for the entire human being. Some scholars believe that ginseng came into use in China around 3000 B.C. The first written mention of ginseng was in 33 B.C., in a treatise called *Interpretations of Creatures.*

The Chinese system of medicine developed through millennia of records kept by court physicians, whose livelihood and favor depended upon keeping the emperor in prime health, rather than merely treating health problems that had arisen. Generation after generation, they kept and passed down records of what failed, and what worked. In the traditional Chinese system, herbs and other medicines were classified in three categories. The lowest grade merely expels disease, the middle grade corrects imbalances in bodily systems, while the highest grade, including ginseng, strengthens the life force of the whole organism. For the Chinese, the Man-Root was treasured like no other medicine.

But somehow, raising ginseng was never quite the sure ticket to riches that Kirk and other promoters made it out to be. For every reader who wrote in crowing over the handsome price he'd just received for his crop, another letter would arrive complaining bitterly over the miserable deal he'd been forced to accept. For one thing, the market was extremely volatile, with prices for ginseng soaring and plummeting in a matter of weeks, for reasons no one could ever quite explain. At times, there were obvious causes: disease destroying a year's harvest and forcing the price up, or war playing havoc with the exchange rates of Chinese currency and wiping out demand. At other times, it seemed the market was simply rigged against the growers.

In the January 1917 issue, W. M. Penrod of Minnesota writes about how his family had spent six years raising their first crop of ginseng—he, his wife, and his seven daughters aged eleven to twenty-six, "all at home." After digging and processing their crop (four hundred pounds of dry roots, a laborious task), they sent samples to twenty-two dealers in Minneapolis to sell. Their samples drew insultingly low price offers of $1.25 to $3.50 a pound, "with various faults. 'Too boney,' 'washed too clean' and the Lord only knows what else." Finally, they accepted an offer of $3.50 a pound, and took their crop straightaway to Minneapolis. When they arrived, they found the company "had a wire that ginseng had taken a slump and they could not take it at any price. You say, how did we feel? Guess!"

Kirk tried to do what he could. He ran reports from around the country, about who had sold what quantity of roots, where, and for how much. He printed long discussions of what kinds of roots the Chinese were looking for, and the merits of wild versus cultivated. But basically, the problem was secrecy. Chinese buyers refused to state directly what they wanted, hoping to buy the choice and the ordinary for the same low price. The big ginseng dealers in New York and other major cities, meanwhile, classified ginseng into myriad grades, using criteria which they would not divulge, saying that they were judging it with the same system the Chinese used. Needless to say, few roots met the standards for the highest price. Some claimed the Chinese wanted the roots dark and gnarled, others said they wanted them thick and fleshy. Growers debated whether there could actually be such a thing as trends in a medicine thousands of years old.

Just marketing one's crop could be a lengthy exercise in frustration. Most growers were located in rural areas far from the East Coast, and at the time monopolistic railway companies were charging extortionate shipping rates, based not on distance but on their control of markets (something like the structure of airfares today). In order to keep charges down, ginseng growers typically sent samples of their roots around to big-city buyers, and then shipped the entire crop off to the firm that made the best offer.

Too often, they were swindled. A typical tactic was for the buyer to claim that the sample had not been representative of the crop, and then send a much lower payment than had been agreed upon. With their whole crop in the dealer's hands, farmers had little recourse. One Michigan grower complained to his congressman that he had shipped his harvest to a New York dealer who had offered him a price of $11 a pound based on a sample. The dealer then claimed the quality of the crop was poor and refused to pay more than $2.50. In the time-honored tradition, the congressman promised to look into it.

Another problem that plagued growers was theft. From the very beginning, ginseng's high value per pound made it a magnet for the unscrupulous. In one of the first letters received by the magazine, Ben Hoyt of Wisconsin reported: "Some one [*sic*] got into my ginseng garden and stole a lot of my first and best ginseng or I would have had more to sell this year." Year in and year out, whatever the price, there were reports of ginseng disappearing just before harvest time. One month a photo was printed depicting "Hunting Seng Garden Thieves With Bloodhounds in the Wilds of Wisconsin."

Finally, Penn Kirk took up the topic, with his usual bluntness: "Recent reports from some sections mention thefts of different amounts of ginseng stolen. One writer suggests that state laws be made to stop this in an indefinite measure. But a thief remains a thief in spite of state laws. He also suggests that trained bloodhounds could be profitably used to run down the thieves. That in a measure makes the ginseng grower his own policeman. So far, so good. But if we have to watch our holdings against thieves I would suggest that a well trained shotgun would be a splendid policeman to save a crop from being mutilated in that way. A suggestion of hunting bird-shot instead of other people's ginseng would not need repeating to prevent it. I am reminded of a fellow in Kansas who had a cow that was clandestinely milked at night by a thief. He inserted a local in his weekly paper that his cow was being milked that way and he thereby gave notice that he was liable to shoot at his cow in the dark some night. Well, it had the striking effect of stopping the clandestine milking and it did not hurt the feelings of the cow either. But if a thief thought that you were liable some night to shoot at a noise in your ginseng garden, he would not likely fool around with it, and you'd not have to shoot either. Use a little printer's ink and try it."

Violence was no idle threat. A rival ginseng publication, the journal *Special Crops,* carried a harrowing report from a grower named L. J. Wilson entitled "How One Ginseng Thief Got His Reward." Wilson, who lived in the mountains of Virginia, had bought a plot of land four miles from his home and transplanted a large number of ginseng plants into the new garden, surrounding it with a rail fence six feet high. After he had been robbed on four separate occasions, his anger got the better of him, and he set up a system of shotguns and trip wires to protect his field, "so anyone coming into contact with or cutting the wire with pliers would fire the gun that he was in front of and will get shot every time."

In July 1908, two brothers named Millard and Rue Collins left their home about thirty miles away. According to Wilson, they "told their people they were going to Tennessee and were coming by to rob my ginseng garden and take a load of it to Tennessee to pay off a fine Millard had against him in that state (this being his former home) so that he could move his wife back there." When the brothers climbed over the fence of Wilson's ginseng field, the guns went off, just as intended. Millard was killed instantly, and his brother fled, wounded. From Tennessee, Rue wrote to his family telling what had happened, and Millard's wife came in search of her husband a week later, the same day that Wilson found the body. A coroner's jury later exonerated Wilson, then dug a hole near the gate and tipped the badly decomposed body into it. "His people being a sorry low-down class they have never moved his remains, and he sleeps on there in peace, and my sleep is more peaceful too, for I'm not afraid of my ginseng garden being robbed any more . . . my guns still hold their position with their eyes wide open day and night," he concluded.

But most ginseng growers hesitated to go this far. The issue was never resolved, in Kirk's *Ginseng Journal* or elsewhere. How do you protect a tremendously valuable crop that grows best in isolated places and takes years to mature? Nearly a century later, people are still struggling.

It seems logical that cultivators of ginseng would try to band together to address their common problems. As early as 1903, state associations of growers were forming, first in Michigan, then across the ginseng-growing region.

Their annual summer meetings were both a forum for exchanging agricultural information and an excuse for a good party. The first day was generally devoted to papers and official business, the second to visiting ginseng "gardens" in the neighborhood and socializing. One invitation from the Michigan State Ginseng Growers' Association, addressed to "Brother Ginseng Grower," announced: "A pot-luck dinner will be served on Mr. Pierce's lawn, so be sure to bring your wife, sister, or sweetheart along and a well filled basket and enjoy the day with Mr. Pierce, who is a progressive grower and has one of the finest gardens in the State."

The state associations also hired scientists to do research on the causes and cures of ginseng diseases that were beginning to spread through their fields. In the wild, ginseng grows in widely scattered patches, so it is difficult for disease organisms to take hold, and they spread only with difficulty. When the plants are cultivated close together, they are much more prone to pests of all kinds, and a blight had nearly wiped out the harvest in 1904. Members of the state associations paid extra assessments for research on these problems, and probably considered the money well spent.

However, Penn Kirk, as usual, had tart observations to make: "These state associations," he wrote, "took it into their processes to hire these expert plant ologists and do the experimenting at great expense among themselves. They added to their expenses what they called a disease fund, to be paid to experts as expenses to spend their summer vacations here and there as they drew their regular salaries at the 'experting' in various places." The expense caused a lot of the smaller growers to drop their membership. On this basis, Kirk made a good argument for the establishment of a national organization to undertake ginseng research and provide strength through numbers in dealing with the market.

In an era in which long-distance transportation was invariably difficult, expensive, or both, it took a long time for a national group to be formed. But by 1916, the *Ginseng Journal* was printing letters from readers clamoring for the organization of such a group. Wrote J. Marchel, of Dancy, Wisconsin, "I for one am real anxious to see the growers get together, that is organize into one body instead of as at present each state by itself with a handful of members, and handicapped so they cannot do anything with the market situation."

In July of that year, the *Journal* published a lengthy paper from J. H. Koehler, one of the largest growers in Wisconsin, proposing the establishment of a "Co-Operative Ginseng Growers Buying and Selling Association." Growers would buy stock at $10 a share in the association, which would maintain a warehouse in New York for drying and storing the roots. After grading the roots by openly published standards, the association would market them to obtain the most favorable price. Members would be assessed a 5 percent commission for storage and marketing of their roots.

The plan was worked out to the last detail, and Penn Kirk supported it most enthusiastically. He offered the services of his journal as the official organ for the association, and urged a meeting to convene as soon as possible to put the plan into motion. In October 1916, delegates from Wisconsin, Illinois, Michigan, Indiana, Tennessee, and Missouri met in Wausau, Wisconsin. There they laid the groundwork for an association to control the supply of this ancient root so prized in China.

By December, it's clear from the pages of the *Journal* that Kirk had already had a falling-out with the organization. "As a person, Mr. Kirk was sort of a waspy individual, and his ire could be easily aroused," wrote Val Hardacre, another ginseng pioneer. "This made him a party to most of the controversial disputes that beset the ginseng business in those times."

Kirk appears to have been infuriated by the association's refusal to make his *Journal* their official magazine. True, its circulation was already healthy and growing, but the additional subscribers and advertisers would have meant a hefty increase in his income. For whatever reason, the officers of the National Association of Ginseng Growers stopped sending him their news reports to publish, and he turned his wrath broadside against them. The $10 shares were a "grab," the 5 percent commission was extortion. The large growers had too much say in the organization, and rival publications (including his perennial competitor *Special Crops*) contained too little news and too many verbatim reprints of free U.S. government agricultural bulletins.

The man could be maddeningly two-faced. To a reader charging that he had been too vociferous in his dislike of *Special Crops* editor Charles Goodspeed, Kirk replied, "What ever got that into your noodle? We met Mr. Goodspeed last winter for the first time in life, and found but a few days difference in our ages with a slight difference in his favor for beauty and youth. He smokes cigars and I don't. I took his paper and paid a dollar a year for it over many years, and quit because I could never coquet [*sic*] him to pay 50 cents for a trial year for the Journal." Sometime later, the two rivals managed to agree on exchanging complimentary subscriptions.

Another opponent, Emer L. Wilder of the National Association, didn't fare as well in the columns of the *Journal*. In a "Little Explanation," Kirk lobbed the thesaurus at him, calling him "this cynical carping, censorious, satirical, sarcastic, captious, snarling, snappish, warpish [*sic*], pettish, petulant, peevish, touchy, testy, crusty, churlish, crabbed, cross, morose, surly, ill-tempered, ill-natured pessimist." Wilder's response was not recorded.

Kirk continued to attend meetings of the state associations, though his health was increasingly poor. "Just after the July Journal went out," he reported, "I was smitten flat with heart leakage and dropsy." Nonetheless he went on with a long, rambling description of his traveling for thirty hours on a Lake Michigan steamer to get to the Michigan state meeting in Frankfort. He claimed that, even though his family was horrified at his traveling so far while so ill, traveling by boat had miraculous restorative powers. At the meeting, he joined in the discussion of cheaper methods of shading, whether letting weeds grow in the patch produced better ginseng through the "survival of the fittest" principle, and the threat posed by low prices of "foreign" (Korean) ginseng.

But gradually the pages of the *Journal* began to fill themselves with large engravings of miscellaneous herb plants, and an eight-page list of salable herbs, repeated identically every month. No issue appeared for December 1918, and the columns of reports from his readers were gradually replaced by pointers on selling furs, "Valuable Things In Scripture," an essay on a dogwood cane. The National Association of Ginseng Growers had become defunct, or irrelevant. As a premium to attract new subscribers, Kirk now offered a copy of a book titled *The Herbalist and Herb Doctor.* The date stamp on the back cover of the library copies I read shows that the magazine was arriving later and later. Where the earlier issues invariably came during the first week of the month on the cover, now they weren't being received until the very end of the month.

The final issue appeared in March 1919. After a hiatus of several months, "The Old Log Cabin" was back. It began: "There are said to be fish in the Mammoth Cave of Kentucky with no eyes . . ." and meanders off into a random string of anecdotes linked only by their having something to do with eyes. An uncharacteristic wistfulness weaves through the essay: "Some things are sad in youth that are beautiful in old age. The accidents of childhood that are healed were beautiful when old age overtook us. They made us glad to be overcomers by whatever help it might have been."

There are letters from growers about dishonest buyers, about dealing with diseases and a method for rigging up a coal-fired boiler to run an irrigation pump. And the last issue ends with a list of herbal cures for asthma (bark of wild plum), "sick and nervous headache" (blue skullcap and catnip), and rheumatism (tansy), and a letter complaining that recent issues have been coming late—"I had begun to wonder what the matter might be around Arrowsmith." Kirk printed it, but, uncharacteristically, didn't respond.

Within a few months of producing the final issue of the *Ginseng Journal and Golden Seal Bulletin*, Penn Kirk was dead, "of a heart condition," according to Val Hardacre. In his lifetime, he had seen the destruction of the forests that abounded in wild ginseng, and the rise of a singular industry that attempted to cultivate the plant. Now the National Association of Ginseng Growers had already faded from the scene, and many of the state ginseng organizations ceased to meet. The boom in American ginseng was fading; many growers were giving up.

Penn Kirk, who printed seemingly every last piece of mail ever received by his *Journal*, left testimonials to some of the bitterness and disillusionment they felt. Wrote T. N. Kincaid: "With corn and hogs I may lose one year and make a profit next year. With ginseng if I have a few bad markets I have lost all, for one has very few market years in a lifetime, unless he grows it so that he can market some each year, and unfortunately I have been unable to do this. I am now 54 years old. The work is hard and help scarce, so with great regret I am quitting, when I can sell out."

Another letter was even more poignant. "Dear Sir," it read, "Kindly strike my name from your subscription list. The ginseng business has been a dream with us, and an expensive one at that. So I do not want to be reminded of it any more. Yours truly, J. A. Burkee."

TWO

OUT SANGIN'

"Don't step on it!" Jo Wolf warns me, too late. "Oh. Oh, well." Then she grins. "I've done that plenty!"

The first wild ginseng plant I've ever seen is peeking out from beneath the muddy Vibram sole of my hiking boot. It is, or was, a baby—a three-leaf. That's why I didn't spot it here on the misty hillside until too late. I hastily lift my foot and try to resurrect the poor thing. Its bright green leaves are tattered, but the stem bounds up tough and wiry. It will probably be okay.

A hundred and fifty years after the young Penn Kirk noticed it was getting harder to find, and a century after writers began predicting the extinction of wild American ginseng, we're out in the Kentucky woods on a rainy June morning, counting to see just how many plants are there. I may or may not have reduced that total by one, but Jo doesn't seem worried. A lot of plants have come and gone in the twenty-six years that she's been doing this.

In 1977, the United States signed a treaty called the Convention on International Trade in Endangered Species, or CITES. In doing so, it agreed to halt international trade in the country's endangered plant and animal species—this treaty was what shut down the global trade in African ivory. In order to

---

Previous page: Panax quinquefolius, American ginseng, growing on the shady forest floor in the eastern Kentucky mountains.

legally export wild ginseng, the United States must show that the harvest doesn't threaten the survival of the species. And in order to show that its survival isn't threatened, someone must go out and count the plants. In the Commonwealth of Kentucky, that someone is Jo Wolf.

Every year, she spends two summer months traveling the state, checking and recording the status of ginseng at nearly three hundred sites. She's not trying to count every last plant in Kentucky—rather, she is recording changes over time in the different locations she surveys. Once every five years, she revisits each site, in a cycle that begins in the eastern mountains and rolls slowly westward across more than four hundred miles, from the Appalachian peaks to the low banks of the Mississippi River.

Ginseng grows in every one of the state's 120 counties, including such urbanized areas as Louisville and Lexington, and Jo knows exactly where to look for it. Some of her research sites are on public land, such as state parks and national forests, while others are on private farms or homeplaces. Over the years she has built up relationships of trust with the rural landowners whose property she visits, and by now they seem remarkably nonchalant about letting an outsider prowl around in search of a wild plant that's worth a small fortune.

Short, energetic, and blonde, today Jo is wearing bleached jeans and a blue work shirt that seem unnaturally clean for someone who spends her days tramping the woods. Her leather boots, though, are battered with scars, their heels ground down. Our first site today is in the middle of a stretch of public land in the mountains of eastern Kentucky. We're accompanied by Ray Combs (his name has been changed), a ranger who tends to this chunk of forest. Jo and Ray make their sure-footed way along the hillsides, over wet fallen leaves and slick downed branches, while I stumble along in their wake, trying mightily not to go tumbling down the hill.

Though Jo grew up in Indiana and talks like a Midwesterner,

bits of Kentucky cling to her speech—"flowers" has one syllable and rhymes with "stars," and "couldn't" rhymes with "puttin'." Ray, on the other hand, speaks in the slow, pure twang of the mountains.

He's a stocky older man, who knows these woods intimately, and loves them. Pointing out an enormous poplar tree to me, he compares it in detail with another poplar mountainsides away. He seems to be on personal terms with every individual tree in his charge, and the part of his job he loves best is leading guided hikes through the woods. We make our way along the muddy trail, pausing now and then for a breather. "One woman came here, she did the whole hike, but all she really wanted to see was the pink lady's slipper. She was from Somerset, and she called over here before she came to be sure when the pink lady's slipper would be blooming. Me, I think the yellow lady's slipper is even prettier."

The first step in doing the ginseng census is to locate the exact site where the count was taken last time, five years ago. As we near the spot, Jo searches for familiar landmarks.

"Did we go out as far as that white oak there?" asks Ray.

"I remember being kind of close," says Jo. "Because this is the hillside. I *hope.*"

I ask her how on earth she ever managed to find a particular hillside again five years later, before the advent of GPS (Global Positioning System) devices.

"Roads, and counting my steps, and angles, and lots of times going with somebody that knows, or remembers, like landowners—because it's their property," she says. "I'm horrible with names, with people's names and faces." This I can believe—she's already decided that Ray's name is Ralph. "But I can remember hill slopes, the way trees look, and just that kind of stuff—maybe because I've had to, because it's a survival thing. I've been doing this for about four years with the GPS. I used to go into the geology library, and from topo maps give it a number, so that every site had a latitude-longitude thing."

Besides, she knows this site well. "I've been here since '78. I came through here three years in a row, and four or five times since then, so this is probably my eighth time. So it's almost like home to me. If I had to do it without Ralph, I would have counted my steps all the way up the trail."

"Do you remember all your sites this well?"

"No, but most of them. My head's full of that kind of stuff. I love music, but I've never been able to talk about it because I can never remember the artists' names, or the names of songs, or any of the words. . . . That's the way it is. So," she announces, "I'm going to drop down a little lower."

She hangs her battered purple daypack from the limb of a young tree to mark the spot she's starting from, switches on her GPS unit, and sets it on the ground among the damp leaves. There are myriad kinds of shiny, wet, bright green plants here, but none of them are ginseng, apart from the unfortunate baby that I stomped flat at the trail's edge.

She and Ray set off in different directions down the hillside, and I follow her as best I can. It's a steep grade deep in soggy leaves, slick after a week's summer rain. Though virtually every square foot of the east Kentucky mountains has been logged at least once, this tract hasn't been touched in a long time, and the trees are massive. We slide and stumble across the hill, straddling our way over mossy toppled trunks so rotten you can shred them with a finger. Once in a very great while, there's the distant sound of a car passing on an unseen road.

"OK, here we go." Much to Jo's relief, she's located the first clump of plants, and she explains to me as she examines and tallies: "There's two babies, and a three-prong, and a two-prong that has a seed pod, too, how 'bout that? Oh, the three-prong's new, it doesn't even have a seed pod. They usually have to be a three-prong before they start doing seeds. . . . Once in a while the two-prongs will put them out, and that's why I've been trying to watch that. And it takes them two years to come up, and then it takes another year or two before they form a two-prong, so that's at least four years. And then they take a couple more years to form three-prong. So I'd say they have to be at least six years old." They're all of four inches tall. "And this guy's got seeds . . . it shows they're healthy and comfortable enough." She marks the tally down on a scrap of paper.

"There's a nice one here, Jo," Ray calls. She hurries over to investigate.

"It sure is!" she pronounces. "But you know, there's no babies around it, and that's what makes me wonder. When they're this big, and you know they've had seeds—it makes me wonder if somebody dug up what was there."

Wild ginseng has two main enemies—development and poachers—and here, poachers are the big threat. Digging wild roots on public land is against the law, but traditionally, in the mountains, wild plants belong to nobody except the finder. It's a tough attitude to change, and Ray says he's encountered this all too often himself. Once, he came across two young men hard at work with their ginseng hoes who didn't even bother to stop when he approached.

"I asked them what they was doing, and they said, 'We're in here ginsengin'.' And I said, you know you're on public posted land. And they said, 'So what?'"—he imitates their baffled tone. "So I said, 'How come you guys came out here?'—they was from Hazard, about fifty miles away. And they said, 'Our moms was up here, and they saw the ginseng, and they told us about it.'"

But, Ray says, legal or illegal, it's no way to get rich quick. "It takes roughly two pounds of ginseng to dry out to be one pound. It would take a long time. I think I would prefer getting me a minimum-wage job. It's really hard work. You get in these woods walking on the terrain we've been walking, and you do that for eight, ten hours, you've done a lot of work! And one thing about it, you're going to have to be in real good shape to do it."

I can vouch for that, already. So why do people still go out sanging?

"A whole lot of people really enjoy it, to be honest. Really enjoy digging it, and finding it, particularly. Plus it brings in extra money. I don't know exactly what it's bringing this year, but that right there determines really just how many people dig ginseng, because when it gets up to four or five hundred dollars a pound, then you really got people out there trying to dig it." Ray looks as disgusted as he ever gets.

As we crisscross the hill, Jo tells me the tale of how she got this singular job, her eyes never leaving the ground. Newly divorced, she had met the man who would later become her husband. He was in the process of starting a carpentry shop in the small town of Monticello, Kentucky. Jo was working as an assistant at the University of Kentucky's horticulture lab. "I had a bachelor's degree in chemistry and anthropology, never took a botany course. Just one biology course, and I was really lousy at it. And I was working in the cytochemistry lab. Part of my duties was to go to the library and take notes for the professor, and see if it was worth it for him to go in and delve into it. And I ran experiments in the lab—we did a lot of experiments with chrysanthemums, the spider kind. We would freeze-dry the roots. . . ." She stops to nudge aside a plant with the toe of her boot, but the leaves peeking from underneath are Something Else.

"Anyway, in the process of going to the library and looking up stuff, I ran into ginseng—little blips in relation to cytokinens, which is a substance that determines whether a plant is gonna die or not. I just kept finding these little blips of information about ginseng, and I thought: 'Oh, *man!*'" She says it in a tone of astonishment. "It was probably just all folklore—but still it always interested me.

"And I thought, I could really get into this. Even if I have to go back to school. My father was *horrified.* I was the only child, of course, divorced . . . and he was trying to get me to go back to school in physics! So I talked to Dr. Roberts, who was the head of Extension Services then. He was giving these lectures on the monetary value of ginseng, because he was trying to cultivate it. I'd figured out I didn't like working in the lab, and being in Lexington—I'd been there for a year. And by that time I was kind of falling in love with Ed, and wanted to move to Monticello. So I asked Dr. Roberts if he was going to be doing any research."

He wasn't. But only a few days later he was contacted by a representative of the U.S. Fish and Wildlife Service. Because the survival of ginseng was so tenuous, it was listed as a threatened species—one that needed careful follow-up. At the same time, it was an extremely valuable export, a multi-million-dollar industry. "And Dr. Roberts told me, we're thinking about doing a grant and keeping an eye on wild ginseng, but we're not sure how to go about it. And that was how I just kind of walked into this." She was a city girl who had never even seen a ginseng plant in the wild.

Over the next few years, she cultivated a network of contacts across Kentucky who scouted out ginseng sites for her. She enlisted the help of county agricultural extension agents, who were familiar with the terrain and could provide a personal introduction to mistrustful landowners. Without local people serving as go-betweens, she never could have done her counts—ginseng is just too valuable to discuss with strangers. And she worked out her method: choosing a plot of land a hundred paces by a hundred paces, noting down data about the location, and then recording the number and ages of the ginseng plants she found. She married Ed, moved to his place in the country, and raised a family, all while spending two months every summer roaming the forests.

"I just lucked into it," she says. "Do you believe in fate? It wasn't something they made for me, it wasn't something that I expected, it was something that *happened.* And to have the nerve to ask for it! I'm not an aggressive person. I never think I even asked for anything like that before. It was too cool. And I remember thinking before, 'Why doesn't somebody just pay me to walk in the woods? This is the most wonderful thing!' I almost never walked in the woods, but when I did, I loved it. I feel very fortunate."

All the while, we're walking, she's counting, and they're both looking for signs of illicit digging. At one point she pauses on a steep slope and bends over to brush away dead leaves. Frowning, she scrutinizes a hole in the ground, just a couple inches wide.

"Hmm. It could even be deer." Then her feet shoot out from under her and she tumbles downhill, landing on her rear. "Well, I just enhanced it tremendously! I fell on the dig spot. . . . Ralph, I don't know if you're going to be able to tell . . . I just shined it up some! See?"

Ray pokes at it carefully with a stick, then shakes his head. "I don't believe it's a digger—'cause a digger would leave a print from a ginseng hoe. It would be a pretty good-size hole, because they'd dig around it to get all the root."

It amazes me to think that this plant, rare as it is, can still be found in every county across the state. But over the years, Jo has "lost" a number of her sites. Woodland has been strip-mined, bulldozed, turned into cow pasture, built into subdivisions. "I lost sites to trailers, to people wanting to set up homes in the middle of nowhere," she complains. "With all the hillsides there are, it seems they could have used many others, but it seems like wherever my ginseng spot is, is where they end up going."

I find it hard to believe that she even counts ginseng in Jefferson County, where Louisville sprawls along the busy Ohio River, but yes, she has a site there. "The last time I was back there, in '97, I finally got there, and the people said, 'Yeah, I don't know anything about it, we logged that a while back,' and so I had no hope. Well, the ginseng was still doing great, but the whole time I was there, I kept thinking—what's that noise? I've heard that noise before. I was so into the plants it was totally in the back of my mind. It was a *beep, beep, beep*. . . . Well, it was heavy equipment backing up! They were building, just outside the trees. Anyway, that's my one spot in Jefferson County."

The final tally for this site: thirty-two plants, just like five years ago. But they're not the same plants, and Jo isn't happy. "Even though we have the same number of plants, it's younger. The older stuff has been dug. And unless they leave it alone, it will just keep getting younger, to the point where it will get wiped out. It won't be able to produce seeds."

We're off to "do" another county after we finish here, and

Ray knows Roger Sizemore (his name has been changed), the landowner we'll be visiting. He's also heard that there's been some trouble there. "Do you think Roger got cleaned out, too?" he asks, concerned.

"I'd hate that for him," says Jo. "I'd like to think that sometimes people—specially somebody that's into ginseng that much—I always hope that he dug it himself and didn't want to tell me he dug it. But as much as he's into it, I doubt that he did it."

"He's a nice guy, Roger is, I like him. He's got some good kids, too," says Ray.

But Jo is preoccupied with her GPS unit, which she has belatedly retrieved from its resting place in the wet leaves. Unless she can pinpoint the precise location of the patch, her data will be useless. "It's probably just too cloudy," she says.

Ray peers at the device and agrees. "Can't get no satellites," he diagnoses.

Then Jo laughs. "It's probably just ticked off I left it on so long!"

At last she coaxes it into providing a reading, and I watch over her shoulder as she completes her record sheet. On it, she has noted the type of soil, depth of soil, compass facing, and a long list of the other plants and trees near the ginseng. "Then the GPS and I do the slope degree with the thing-a-mometer—what do you measure angles with? It always used to work great, but now I think it's been in too many rainstorms. It just kind of goes someplace and sticks."

Finally, we load her gear back in the trunk of her little red Toyota, and she takes her leave of Ray.

"You've been doing this since 1978?" he marvels. "I just now figured—we're getting old!" We all laugh.

"I've been doing it twenty-six years," she says, "and each year I never know whether I'll get the grant or not."

"And I've been here thirty-two years," says Ray. "Started 1973."

She shakes her head. "It's weird to me to think I've even been around that long."

"What are they gonna do when me and you both go, Jo? You know . . . I'm the last one here. Dr. Powell, he's semi-retired, so when he goes and I go, then there ain't gonna be nobody."

They shake hands, promising they'll see each other again in five years.

Mr. Sizemore's place is down a gravel road off another gravel road off a blacktop road that narrows to a single lane where the monstrous coal trucks have torn away the edge of the pavement, leaving a steep drop into wooded nothingness. It's not a long drive, but it sure feels like it.

He lives in a tidy brick ranch house with flowerbeds and a flock of goats out front. Two black horses graze in a field next to the house. He and his wife come out to chat with Jo for a moment before we plunge into his woodlot. He's a soft-spoken older man, retired now, with plenty of time to ponder the world and its wicked ways. Jo has been there counting ginseng a number of times before, and he doesn't seem particularly bothered that a mud-splattered writer with a tape recorder is dogging her footsteps. Jo promises to come back and sit a spell after she finishes counting, and we hop cautiously over the electric fence, the watchful horses eyeing us.

"He had 865 plants in here," she tells me. "I don't think I'm going to get anything close to that."

But my eyes have become attuned to the flat clusters of leaves, and now I can see that there's ginseng everywhere—a few plants nearly two feet tall, a handful of "teenagers," and lots and lots of babies, scattered all across the rocky hillside. Jo counts under her breath, turning and scanning all around her, while carrying on a running conversation. Originally, only wild ginseng grew in this patch of woods, but Mr. Sizemore started collecting and replanting the seeds, then brought in new seed from outside to augment the supply. By now it's so interbred it's hard to say whether it's wild, or cultivated, or what. If this profusion of plants is just what the thieves left behind, they must have found themselves a gold mine here behind the house.

Haven't you ever been scared, I ask her, out roaming the woods all alone?

She nods. "I got lost once, in Perry County. That's my main fears: bad ticks, and being lost in Perry County. . . . It was just before dark, and I got turned around.

"It had never happened to me before. I started running. And I thought, Ohh . . . this is what you're *not* supposed to do. And so I slowed down, and I sat there, and thought, 'I'm not even sure where I am.' I'd driven back to the end of a road, it was an oil-well road, and then walked beyond that, for quite a while, and then went down a hillside and up the other hillside. It was where I had a ginseng spot, where I'd been twice before that, so I didn't have any reason to think of getting lost. And when I went back up the hillside, it wasn't the same. I don't know what happened. I still don't know. But I just thought, well, you don't want to spend the night in the woods. Just the thought of something crawling on me was really creepy. So I thought, just anchor your spot, so you can come back to your spot. . . .

"And then I looked up on the hill, and there was a beer can there. And I thought, if there's a beer can here, surely I can just keep following that trail out. And that worked. They wouldn't have started looking for me if I was lost. . . . And you know what the weirdest thing is? When I went back to that spot, the next time, it had been leveled. They had mined it, it had been reclaimed, and a trailer was sitting on top of it. And that's where I was lost! Have you ever seen a picture of heaven and hell, like they put in the Bible? The grass had just been planted and the trailer was brand new. This was somebody's idea of heaven, but to me it looked just like hell."

But didn't she ever feel nervous, running around the woods alone?

It's not the being alone that scares her, she tells me, it's the people. "The only other time I felt scared was when I ran into some people digging ginseng, and I kinda tiptoed around them. I was way back in the middle of nowhere, and I knew it was just me versus them. I walked right on by, silently. They were up on a hillside, and they were digging. They had hoes, and pillowcases. . . . That was in Knox County. Barbourville."

She pauses. Maybe she's had more scary times than she realized. "And then one time I was in western Kentucky, it was right around Edmondton. . . . That evening when I walked back out, I had to walk through a field, and I got in my car. I was driving home, and it was like four hours, and I looked down at my arm—I had on a white shirt, and all these little critters were walking up it. I had walked through a tick—FARM, of some kind! There were little bitty seed ticks running all up and down my arm. I can't believe I didn't have a wreck. I ran out in the middle of this field, took all my clothes off—I felt so creepy! It was the *worst* ride home. So, ticks and people are my biggest fears when I'm out in the woods. I know that's not what you're supposed to say." But then she's on to more cheerful topics. "Hey, look at that sneaky plant—there's one right under there!"

As we crisscross Mr. Sizemore's woods, Jo gives me lessons on the inscrutable growing habits of the ginseng plants, her observations of many years. For one thing, she says, the books are all wrong when they claim that the plant follows a set pattern. Often the plants don't come up for a year or more—they just stay in the ground. New growth comes from a shoot on the side of the root, and if that is knocked off or damaged, the plant stays dormant while it heals. "It takes a year for it to think about it," she says.

Later, I finally remember to ask Jo a question that's been nagging me: "What do you do the rest of the year?"

"I make hats," she says.

"*Hats?*"

"Yeah. And Ed's a carpenter, so he always needs some kind of finish work. If we're lucky and he's working, I do varnishing, and staining and sanding, stuff like that, and that works out good. And I make hats. Cloth hats—I had some in the Upstairs Gallery for a while, in Berea. Roll-ups and cat hats and all kinds of hats. . . . Look, there's a lot of little two-prongers here."

In the end, Jo tallies up 175 ginseng plants on the Sizemores' property, most of them very small. Thieves had gotten the other 690—dug them up without replanting any of the seeds.

We sit down on Mr. Sizemore's cool back porch to visit, while his wife tries to talk some sense into their dog, a huge, unruly black Lab puppy. Jo perches atop the retaining wall, swinging her feet and banging her boot heels against the bricks.

"Do you know anything about keeping shrews away from your ginseng?" Mr. Sizemore asks Jo anxiously, but alas, she doesn't, so he tells us about his last attempt to deal with them. He was fed up with shrews eating all the berries off his ginseng and keeping the plants from seeding, so he decided to pay a guy $20 to catch him a blacksnake. But the man brought it while Sizemore was on vacation and for some mysterious reason left it in a cardboard box up on his roof—why, he doesn't reveal. "It was about all burnt up," he tells us, disgusted. "But I let it out of the box, and ZIP! It was gone, just like that." He saw it around the property every so often after that—you have to bring snakes from far away, he explains, because if you get them from just a mile or two away they'll take off straight back home again. This one was imported from twelve miles down the road. And it seemed to be doing its job, until one day his son ran over it with the lawn mower. "I came out and found just a big pile of meat on the lawn."

But what about the poachers?

He shrugs, resigned. One day, he was out shooting at a turtle in his pond—he loved fishing, and he owned a little ol' backhoe, and so he dug himself a nice fishing pond, right near the house. But the problem was, big snapping turtles kept coming and eating all his fish out of the pond. And he tried to catch them to take them down to the river to live, but you just can't catch a snapping turtle, so one day he decided there was only one thing to be done about it. And he'd no sooner shot the first turtle than he heard people running away up through the woods, behind his house. They were ginseng thieves, and they figured he was shooting at them. He went into his woods and found all the holes where plants used to be.

Jo commiserates with him, and tells him that he has lots of fine-looking baby plants back there. She'll be back in five years' time, to see how they're doing. But Mr. Sizemore doesn't know whether he's going to keep on with the ginseng, given what's happened to him.

"It kindly got me disheartted," he says softly.

Then he excuses himself. It's time for him and his wife to leave for their Bible study.

I've been given the phone number of a man in eastern Kentucky who I'm told knows a lot about digging wild 'sang, but it's not so easy to get hold of him. I keep calling. This time, just when I'm about to hang up, someone finally picks up the phone, and a gruff female voice says hello.

"Hello, is Mr. Slone there?"

"Nope."

"Well . . . ," I say lamely, "I guess I'll try again later."

"Suit yourself." Click.

Jo Wolf is off to another county the next day, and hoping the weather holds. I head down the highway to the town of Jackson, to visit the Breathitt County Public Library.

Breathitt County lies at the heart of Kentucky's Eastern Coal Field. It's not the poorest county in the state, but it's not the richest, either—the median household income in 1997 was $18,404, next to a state median of $31,730. When the railroads came here in 1888, they opened the county to mining and logging. By the 1920s, it had been stripped of the most valuable timber; by the 1930s,what had once been rich agricultural land had lost most of its topsoil. Now all that's left is the coal.

Steve Bowling, the county's head librarian, is a ginseng expert of a different sort. In his office I take a seat in the visitor's chair, a homey wooden rocker, while a police radio chatters away to itself on a desk. It doesn't surprise me to learn that Bowling serves on the local rescue squad. Tall, dark-haired, and athletic, he looks like just the sort of person you'd be desperately glad to see when you're in big trouble.

Before taking the library position, Bowling worked as editor of the town's weekly newspaper. Kentucky still has a phenomenal number of small local papers, and the best of them are scrappy and partisan, ardent connoisseurs of local history and culture. As editor of the *Jackson Times,* Bowling was also the paper's main reporter and feature writer, and he wrote close to two thousand articles on local issues—including ginseng.

People have dug wild ginseng in Breathitt County since the 1840s, he tells me. At first, it was sent down the river on flatboats to market in Frankfort, the capital, in the faraway Bluegrass. In the 1840s, the Legislative Record would list items that were being brought down out of the mountains, as a gauge of the economy there: things like salt, wood, coal, and ginseng. When the railroad finally reached the area in the 1890s, shipping ginseng out of the mountains became easier. By the 1970s, it was pretty well just about gone.

"As long as I can remember," he tells me, "and as long as most people that I've talked to can remember, it's been something that folks used primarily as supplemental income. It was dug by families who were on public assistance." When the economy was doing well, when the coal industry was booming and the mines were hiring, people didn't have the spare time or the energy to go out digging ginseng. But in hard times—which come more often to the Kentucky mountains than to the rest of America—people got out their ginseng hoes again. During the Great Depression, people dug ginseng to help feed their families. World War II brought a measure of prosperity to the mountains, when tremendous quantities of coal were needed in the steel industry, but the region went into a slump again in the 1950s. The War on Poverty in the early 1960s did little to improve matters, in Bowling's opinion—"My personal perception is, if you convince somebody they're poor, then they're gonna act poor. You are what you think." In the early 1970s, the news was so unrelentingly grim, filled with the economic downturn and skyrocketing oil prices, that people headed out to the hills to dig up every last ginseng plant they could find, expecting economic disaster.

"That's been kind of the enduring theme of eastern Kentucky," Bowling tells me. "When the economy goes down, you go to the hills. The economy was bad in the late '90s, and now a fourth round of logging has started. The mountains have always provided everything that people need, whether it's ginseng or coal. Ginseng has always been there for the common man."

Then the phone rings. Speaking with the caller, Bowling fires up his computer and runs a search on a genealogy website, reading off a list of names and dates and asking lots of questions. Before hanging up the phone, he promises to locate a volume with the records of the Deaton family.

"We get 40,000 calls and visits a year at the library regarding genealogy," he says. (The total population of Jackson is 2,490.) "There are families that have been here since 1789." He shows me his record sheet for genealogy inquiries, filled with lines and lines of cross-hatched tallies. "That's for this week." It's Thursday morning.

Bowling was raised right here in Jackson. His family were town folks—"attorneys, politicians, all that good stuff"—but he had one uncle who would go out sangin'. "He dug it any time, he didn't care! If he found it, he dug it. He would string it behind the heating stove on a string, so it would dry out. It would have been about 1984, '85, not that terribly long ago. But the rest of my family never really plundered around out there."

Ginseng has its social aspect as well, he says. "We think of somebody digging it for money, but in the late '70s when the plants grew harder and harder to find, they were used for entertainment purposes. People would locate them, and then they would give a general coordinate to the next group, who they challenged to find this plant. As they grew more remote to find them, it became more of challenge. When you're trying to find two or three specific plants, that's more difficult than finding just any ginseng. That's the only place that I've ever heard anyone mention that being done, but I'm sure if they did it in Breathitt County, they did it somewhere else. . . . Excuse me. . . ."

As he takes another phone call, I look out from his office into the library. It's summer vacation, and clusters of kids are huddled around the computers. Their moms leaf through magazines at a table. A young couple with a baby comes in and asks the woman at the desk about the schedule for walking tours of the town. When he hangs up the phone, I ask whether the library has books about ginseng.

"We do, but we don't. Usually as soon as they come back in they get checked out again. They'll keep them two or three months until you call them up and they'll bring them back, but then they go straight back out again."

According to Bowling, there's still ginseng to be found in Breathitt County, though it may not be as easy as it once was. People tend to overlook such niceties as the ban on digging on public land, or the concept of asking a landowner's permission. "Property ownership here—they always used to say there are two things that will get you killed quicker than any other, and that's property and religion. And in some ways that's true. But when it comes to ginseng diggers, they're seen as kind of harmless folks. Now if they range too close to the house, that may be a different story. But for the most part, they're seen as one man, a little small pick, and a bag, out in the woods. I can never remember there being a trespassing charge because of ginseng, ever. And you have to understand, the vast majority of the property in Breathitt County is absentee owned. So they're not actually living on the property, or it's a coal company. And it's not like they're right down next to somebody's garden, or in their pasture field or anything like that. There are large, large, large tracts of just open land."

Most of the diggers he knows are young men, from twenty-two to thirty-five. After that, the work is just too strenuous for most people. They've learned their craft from the previous generation—where to hunt, and how to look for common plants like goldenseal and cohosh that are often found near ginseng. And most of them know that they must plant the seeds of the plants they harvest, and avoid digging every plant in a patch, so that the species will survive. In addition to digging ginseng in the fall, many of them also dig other wild medicinals, such as may apple, goldenseal, and bloodroot in a cycle through the year, though these don't pay nearly as much.

But there's more to it than just money. "There are folks that seem to get more enjoyment out of how much ground they can cover, rather than finding ginseng. Of course, everybody would like to find a five-prong, but most of what they're concerned about is how much ground they can cover. Ginseng digging is something that's one on one, you vs. nature. You find where nature put it. It bonds you with the property, the community, the territory that you cover. Once you cover it, a true mountain man, that hillside becomes yours."

The police radio on the desk breaks into an outburst so persistent that Bowling finally turns the volume way down. I ask how often he gets called out on emergencies.

"In the last two years we've had five rescues, and four of them have been ginsengers who have been lost. So sometimes nature wins. . . . Even in the age of GPS and of topo maps, sometimes nature still wins." He tells me about a young man they were called out to search for the previous fall, on Rose's Branch of the Kentucky River's South Fork. The rescue squad came out with bloodhounds, who tracked him to a certain point, until the scent vanished. Lost for more than a day, the digger crossed a range of hills and dropped into a hollow, where he finally stumbled across a house. Most of the other people they searched for, being experienced woodsmen, carried emergency kits and extra food, and just built a fire under a rock overhang and settled down for the night, waiting to be found.

It's hard to believe that scenes like this take place in the densely settled eastern United States, but these woods are still wild. "One family, a woman and an eight-year-old boy, they were out digging some bloodroot, almost within sight of their own house, but were so turned around they couldn't get home. We always tell folks, if you get lost, just go downstream, you'll come to something eventually. But they don't. They tend to walk ridges, because they think they can identify something, but the further up a ridge you go, the further you are from civilization. I can't think of any case, in the ten or twelve years that I've been on the rescue squad, that lost people haven't been within two or three miles of their own home. Places that they've walked for years and years. But something draws your attention, you get turned around, or drop over a hill into another creek, and very quickly you're six miles from where you're supposed to be."

One question is still bothering me. If these ginseng diggers are so in tune with nature, then why have they nearly dug the plant to extinction at times? I finally ask, and Bowling turns philosophical.

"There's almost a defeatist attitude here in the mountains," he says, "that if everything is going great, that's bad, because something's about to happen. We've been preached to for years that we're not smart, that we're ignorant, that we're just hillbillies, and there's a thought among some people in eastern Kentucky, they're always waiting for the other shoe to drop. That's kind of the perception with ginseng diggers, too. If the '90s were such an economic boom, everything's rolling along and everybody's making money, you know it's going to end, so we'd better start getting ready for that. Folks were trying to hoard up money, and store away extra emergency cash. It's easy to take a few hours on Saturday and if you find enough ginseng, you could make a hundred, two hundred dollars."

The phone rings again, and it's another long, intricate genealogy question. I gather my things, whisper a thank you, and tiptoe out as Bowling launches in: "Now, the Hensley family . . ."

Try again. This time the woman picks up after five rings.

"Hello, is Mr. Slone there?" I chirp, sounding hopelessly Midwestern.

"Nope."

"Well, when's a good time to call him?"

"Don't matter."

"How about tomorrow?"

"Tomorrow, next month, suit yourself." Click.

Interviewing a 'sang digger, I'm starting to realize, may be harder than I'd figured.

Wild ginseng is understandably a sensitive subject for a lot of people in the mountains. For one thing, ginseng is both extremely valuable and extremely easy to steal—talking to strangers about the ginseng on your land is not unlike explaining just how to steal your goose that lays golden eggs. For another thing, a lot of wild ginseng is harvested illegally, dug on other people's property without permission, or on off-limits public land. Traditionally in the mountains, forest plants were seen as part of the commons, available to anyone who can find it or make use of it. Present-day law takes a different view, and though this doesn't seem to deter people from digging up the plants nearly anywhere they're found, diggers know better than to publicize that. And finally, ginseng digging is, for the most part, an off-the-books, cash-only business. As Steve Bowling points out, many diggers are living on disability benefits from Supplemental Security Income (SSI) or other forms of public assistance, and are afraid of jeopardizing their monthly checks. Three awfully good reasons not to talk about it.

Fortunately, I've been introduced to the work of a woman who has spent years learning from, and about, ginseng diggers. Professor Mary Hufford is the director of the Center for Folklore and Ethnography at the University of Pennsylvania. From 1994 onward, she interviewed scores of people in West Virginia's Coal River Valley about their life in the Appalachian forest.

Hufford's work began as part of a project on the mixed mesophytic forest. Even environmental activists are often astonished to learn that the woodlands of Appalachia are the world's oldest and most diverse temperate forest, home to a staggering two thousand plant species, hundreds of bird and animal species, and uncounted varieties of insects. The term was coined in the 1920s by biologist Lucy Braun, who first studied the region, and work continues today to assess and quantify the diversity of its ecosystem.

As part of the Citizen Science Monitoring Project, Hufford carried out a cultural study looking at the seasonal round of activities in the West Virginia forest, such as collection of wild plants, and interviewed local people about forest health. Ginseng soon developed into a focal point of their discussions. "Ginseng became a window for me onto forest health issues," she says, "because people on Coal River were using it so frequently to illustrate their own observations about forest health. Ethnographic training teaches us to listen to what people are telling us about our questions—I was asking questions about forest health, and they were telling me through stories about their experiences in the woods what the issues are. They were also validating their authority to speak about forest health by telling me about the amount of time they spent in the woods ginsenging, squirrel hunting, combing the ruins, and so on." Within the context of their shared concern for the future of the forest, she would ask people, "Do you have a sang hoe?" and this led on to long conversations about ginseng and how they hunted for it.

Ginsengers in Appalachia dig up the roots using a sang hoe, a metal tool with a wooden handle. You won't find them in any store—they're often made from old miner's picks with the points cut off, or a broom handle with a piece of automobile spring attached. The long handle makes the tool useful as a walking stick, and also as a probe to check bushes for snakes, a common hazard along with bear and yellow jackets. People would get out their homemade hoes to show Hufford, and pose with them for photos. Today the photos, along with recordings of the interviews, form a part of the American Folklife Center collection of the Library of Congress, and can be accessed through the Library's website.

The interviews make fascinating listening. In them, Mary Hufford's light, friendly voice can be heard chatting with people of all ages about all manner of topics, from encounters with rattlesnakes to a splendid tall tale of a ginseng plant so big that its berries rained down like tomatoes on the back of the man who tried to dig it up.

Many ginsengers learned the art from older family members. Among the people Hufford spoke with were three generations of women in one family who were all devoted 'sangers: young Natalie Pettry, her mother, Carla Pettry, and her grandmother, Shelby Cantley Estep. As Hufford interviews the family, their enthusiasm spills over: three voices speak almost simultaneously on the recording, interrupting and finishing each other's sentences. "Once you get it in your mind, it stays there," says one. "If you can't go ginsenging, it totally drives you crazy, it does," the second voice cuts in, and a third continues: "If you love the hills, if you love the mountains."

Some diggers describe ginsenging as akin to fishing: when you go out, you never know what you're going to get, and the thrill of the hunt is just as much a factor as the cash that a big root can bring in. "You get an adrenaline rush when you find one," said Randy Halstead, a ginseng digger and dealer who spoke with Hufford.

It takes years of experience just to know the best places to look, and to spot the reclusive little plants. One of Hufford's interviewees spoke of going out ginsenging with his grandmother, who, though too old to do any digging, pointed out plants everywhere that no one else had noticed. A man named Ernie Scarbrough explained to Hufford, "If you look under the right tree, you might find a stalk of seng. There's trees I go for yet, ginsenging . . . sugar maples and black gum, whenever you can find one. And the hickories. Squirrels is in the hickories, and they eat the ripe ginseng berries. So it makes a lot of ginseng around the hickories." Another sign is the plant he calls "ginseng pointer." Supposedly, the branches of the cohosh plant will point in the direction of nearby ginseng—how nearby is a good question. Other diggers use that name to refer to other plants, like rattlesnake fern, often found growing together with ginseng.

Upon finding a baby plant, diggers look uphill to try to locate the bigger plant that its seed came from. And after digging a plant, responsible ginsengers will scatter its berries nearby, making a mental note to come back in a few years and see what's growing. (Some diggers are not above reinstalling the rootless plant in the hole, to trick the next digger who wanders along.)

In her study "American Ginseng and the Idea of the Commons," Hufford explores the role that ginseng played in the traditional mountain economic system: "Rural populations with uncertain employment," she writes, "have typically relied on gardening, hunting, and gathering for getting through hard times. Over the past decade, processes of gentrification, preservation, and intensified extraction of timber and minerals have eliminated the commons in which communities have for generations exercised fructuary rights."

Historically, in Appalachia, only the hollows and lower slopes were settled and used as farmland, producing crops like corn and tobacco and serving as pastureland. The higher forested areas were seen as a commons, an area available to all which supplied people with wood for fuel and building, an open range for livestock, and a source of plentiful nuts, mushrooms, game, and medicinal herbs—like ginseng.

Today, these forests are being devastated, as "mountaintop removal" strip-mining savages the Appalachian countryside. In the never-ending quest for cheap coal to fire power plants, entire mountains are blown to rubble, then dumped into nearby riverbeds, poisoning the water and leaving miles of blasted moonscapes where nothing will grow. Grudging, half-hearted reclamation efforts result in, at best, a flat shelfland covered with a single species of sparse stalky plants, useful only for golf courses or still more prison sites. Neither are an urgent need in Appalachia. A forest ecosystem that took centuries to develop is annihilated in an afternoon, and ginseng loses yet another patch of its vanishing home.

One last time, I try to reach Mr. Slone. This time I let the phone ring and ring. But no one ever answers.

THREE

SHANG GARDENS

A noisy, dusty street corner in a gritty Korean farming town called Geumsan. To get here, I've traveled three hours on a creaking bus from Seoul, overnighted in the provincial capital, then gotten up at the crack of dawn to ride another bus to meet Mr. Choi. He's supposed to translate for me, and the arrangements were made weeks ago, but it has taken a half dozen cell phone calls to track him down, and it's late in the morning before we finally meet.

Around me, the crowd of locals settles down on the pavement, smack in the middle of the intersection. There are young couples and little kids, a sunburned row of farmers in the Korean version of gimme caps, wizened old women walking with their backs bent to a near–right angle by years of postwar hunger and hauling babies in slings. A grandfather in the seldom-seen traditional clothing of white hemp sits next to his little grandsons—or great-grandsons?—decked out in cartoon character t-shirts. All of them are waiting expectantly.

The ceremony begins in a crash of brass gongs, before I have time to ask Mr. Choi what's happening. Four musicians swathed in purple and white kneel on a mat in the street. To the thunder of hourglass drums and the clatter of brass cymbals, two masked

Previous page: Farmers burn a field following the ginseng harvest in Korea, the first country to cultivate ginseng.

shamans in dark robes sweep slowly into the space, holding aloft smoldering bundles of herbs. Somber women raise tall, white banners painted with Chinese characters, eyes downcast. The jangling bells on the shamans' legs make my scalp crawl in broad daylight.

They stand before the altar, a table heaped with offerings of apples, pears, dates, and rice cakes, all mounded with geometric precision. With their torches, they light a brass brazier of charcoal, and the smoke comes boiling out. Then four men in blue enter, huge swords in their hands. They dance in slow motion, their swords carving through space, and then slip away into the crowd. "This is to purify the space," Mr. Choi says. "Old shaman ritual."

The drums and shuddering gongs pick up their tempo. A young man dances onto the stage, his face a mask of anguish. I can see him pleading, raising his arms to heaven. "His name is Kang," Mr. Choi tells me. "He lived one thousand and five hundred years ago. His father is dead, and his mother is very, very sick. The doctors can't help her. He is a dutiful son. Every day, he prays to the Mountain Spirit." Before the altar, Kang prostrates himself again and again, his forehead hitting the pavement.

Suddenly, the Mountain Spirit appears. He's a towering figure with a long, white beard, swathed in white and capering along atop tall stilts, to the visible delight of grandfather, grandchildren, and everyone else in the audience. Kang, on the contrary, is terrified, but the god raises his hands to reassure him. In them, he holds a plant with a big gold root and bright red berries. It's rather lumpy, stitched and stuffed from cheap fabric, but everyone in this crowd knows what it's supposed to be.

"In his dream," Mr. Choi tells me, "the Mountain Spirit says he should go to the mountain and find the plant with red berries, to make medicine for his mother. And she is cured."

The Mountain Spirit has vanished again, but Kang is dancing his joy and gratitude in front of the altar, holding the ginseng plant in his hands as he passes through the sacred smoke. "This is the legend of how we Koreans grew the first ginseng," says Mr. Choi.

And now the audience is jumping up and crowding around

the altar, and a man is handing out long strips of stiff tissue paper. You light them in the charcoal, make a wish as they burn in your hands, and fan the last smoke and flames onto yourself. A Korean photographer makes me do it all a second time so he can get more shots of the peculiar foreigner.

And suddenly we're back on a street corner in a dusty small town in South Korea, on the final Saturday of the annual Geumsan Ginseng Festival.

Legend aside, it's fairly well accepted that farmers have been growing ginseng in Korea for over a thousand years. Some experts speculate that ginseng may have been cultivated in Korea as early as 11 B.C. Traditionally, it was raised under shades made of straw matting, and by 1400 A.D., Koreans had developed something called the Yang Jik method, which involved using supplemental nutrients to grow ginseng from seed (rather than transplanting young wild plants). Whatever the true dates, no one knows the plant more intimately than the Koreans.

"Would you like to see the ginseng market?" says Mr. Choi. Festival or no festival, people in Geumsan are all business, and the market is going full-tilt. From outside, it looks like just another drab, boxy Korean commercial building—in the aftermath of the Korean War, people learned to build for speed, not beauty, and even in the boom decades they never lost the habit.

But the sight that greets my eyes is startling. Inside, the single huge hall is lined with dozens of white-tiled stalls, and every one of them is mounded, heaped with boxes and cartons and baskets and crates of fresh ginseng root. The sweet, earthy smell is enough to choke me.

And though the wares of every stall look identical, clearly they're not. Some are crowded with elbowing customers; in others, the proprietors (almost all of them hard-faced middle-aged women, *ajummas* in Korean, with black money belts strapped around their waists) stand forlorn. I see a couple of them absent-mindedly whittling off bits of ginseng root and popping them in their mouths.

I step up to one of the stalls. Inside each carton, the roots are neatly arranged, shoulder-to-shoulder—the trailing, threadlike branches make them look like a line of headless hula dancers. "How much are these?" I ask, picking a random tray. About $33 for 750 grams, a pound and a half, comes the answer. "And those?" Fifty dollars. The difference is in their age and shape. Asked about the best quality, the woman brings out a small basket of frilly, twisted roots that have grown for six years. Asked to quote a price, she laughs and refuses.

The air is filled with the sound of tearing plastic bags and women's voices haggling prices. This is the largest ginseng market in Korea, where over 80 percent of the country's crop is sold. On a busy day, up to 150 tons of cultivated ginseng changes hands here—worth 43 billion won a year, over $41 million.

And that's just the raw ginseng. Geumsan's streets are crammed with over a thousand shops and little stalls selling an unfathomable array of medicines made from ginseng and other herbs. Their windows and displays are filled with gaudy red and gold boxes of tea, jars of capsules, and bottles of extracts. In huge decorative apothecary jars, enormous roots are steeping in alcohol. Alongside the ginseng are tubs of dried mushrooms, heaps of bark and twigs, and bundles of dried centipedes, some specimens six inches long. I decide I don't want to know where those come from, or what they cure.

Geumsan has an international ginseng market as well, where processed roots are sold to buyers from China, Asia, and North America. There are offices of wholesalers in ginseng products, and the Ginseng Hotel, where traders spend the night. There's even a ginseng sauna. I decide that, with all the ginseng I'm eating and drinking, I probably don't need to breathe it as well—or soak it in through my pores.

In the afternoon, a van shuttles visitors out to a nearby ginseng farm for some hands-on experience—literally. There, in a field wedged between brilliant green rice paddies and fields of red chilies, you can dig your own ginseng roots, then buy them at a discount.

I've seen these sheds dotted all around Korea, climbing the mountainsides and even crammed inside the Demilitarized Zone at the North Korean border, but this is the first time I've been inside one. Ginseng grows wild in the cool shade of dense forest, and in order to cultivate it, the plant must be protected from direct sunlight. In Korea, this is done by planting it under long, wooden structures roofed with black shade cloth. These sheds are about five feet high in the front, sloping to three feet at the back, meaning that all labor performed inside them must be done by hand—and crouched, or stooping.

It's a warm, sunny day, but inside the sheds it's cool and shady in the dappled light. This farm is on the lower slope of a mountain, and despite two inches of rain the day before, the gray earth is barely damp—a textbook example of well-drained soil. Ginseng is a plant that sends its energy to its roots, and even these healthy green specimens have sparse leaves.

Vanloads of visitors have already been through, leaving several sheds dug up and trampled, but the *ajumma* beams her delight at having foreign guests. Women in the Korean countryside love bright colors, and this one is downright kaleidoscopic: she's wearing an orange, yellow, and green striped blouse, red rubber gloves, black sleeve guards, royal blue pants tucked into hot pink socks, and sky blue rubber shoes. A bright yellow scarf anchors her white hat to her head.

Leading us into a shed, she shows us the harvesting tool. It consists of two foot-long claws of hammered metal, curved back like the horns of a gazelle until they nearly touch the wooden handle. She demonstrates the technique: you grab the plant and stab the claws into the ground along either side of it, then rock the tool back and forth until the root slides free from the loose, sandy soil. A few quick movements and she has the ginseng root uncovered.

Then I try it. My first stroke impales a valuable root neatly on the hook, ruining it. Clearly this is trickier than it looks. The *ajumma* tosses the spoiled plant aside, then shows me exactly where to aim around the next stem. This time I can't get the root to come out. Placing her hands next to mine, the woman throws some muscle into it, slamming the tool into the ground and yanking hard. The plant pops out immediately.

I destroy another four or five plants before I finally get the hang of it, and even then I can see that harvesting requires as much telepathy as strength. You have to be able to divine which direction the root is growing before you aim the hook, in order to avoid piercing it. Around me, families, kids, and a busload of Japanese tourists are all digging away, but no one seems to be having quite as much trouble as I am. The *ajumma* bustles around sorting and bundling roots that the visitors have dug, and ringing up sales. I can't believe that every single root in Korea's multi-million-dollar industry is dug up by a body crouched painfully over this primitive metal hook.

For centuries, Korea was known as the Hermit Kingdom, and foreigners were banned from its territory, except for a handful of Chinese traders. When Japan forced open Korea's trade by imposing an unequal treaty in 1876, the first Westerners to arrive there found ginseng production flourishing. "No one can be in the Far East for many days without hearing of this root and its virtues," wrote English missionary Isabella Bird Bishop in *Korea and Her Neighbors,* an account of her travels there in 1894–1897. "It is one of the most valuable articles which Korea exports, and one great source of its revenue."

Bishop visited a number of ginseng farms in Korea and meticulously recorded details of how it was grown under shelters made of woven reeds, with blinds that could be raised and lowered to control the sunlight. She described the processing method for harvested roots: "Ginseng is steamed for twenty-four hours in large earthen jars over iron pots built into furnaces, and is then partially dried in a room kept at a high temperature by charcoal. The final drying is effected by exposing the roots in elevated flat baskets to the rays of the bright winter sun." After sorting, they were packed in trays, and fourteen of these trays were sealed into a larger basket for export, certified by the Korean government—and worth up to $20,000 at that time ($440,000 in present-day terms).

Despite what she saw, Bishop was not overly impressed with the Korean temperament, and concluded "if industry were lucrative, and the Korean were sure of his earnings, he would be an industrious and even thrifty person."

Since that era, Korean agriculture has changed enormously, but many aspects of ginseng production remain the same. Because the sheds are too low for machinery, the plant is still grown entirely with hand labor, from planting to harvest. And two different types of Korean ginseng, red and white, are still produced. White ginseng, the less valuable, is washed, peeled, and then mechanically dried. Costlier red ginseng is not peeled, and is steamed before drying slowly in the sun. The endless array of products on sale in Geumsan are made from these processed roots.

Back in town, we take in some of the entertainment at the ginseng festival. This afternoon there's a puppet show and an "Insam King" pageant for kids (we watch the talent competition, which includes a boy who sings a song about ginseng and has to start over again four times). The local senior citizens' club plays a traditional game called *yut nori,* which seems to consist of throwing marked wooden sticks into the air and then arguing a lot.

Along the streets, the rows of tent restaurants are cleaning up from lunch and getting ready for the dinner onslaught. Some of the dishes are ginseng-related, but there are also Korean standards, such as *bindaeddeok* (mung bean pancake) and *ddeokbokki* (rice cake in chile sauce). In front of a shop selling elegant hand-carved name seals, big plastic buckets of fresh chicken feet await grilling. This is not the Korea that builds SUVs and camera phones.

Finally, there is the "Nong-Ak Competition." Each day during the ten days of the festival, a nearby village is invited to give a performance of the traditional Farmer's Dance, with a prize awarded on the last day to the best troupe. Today it is Geumseong village's turn, and we quickly discover we've seated ourselves directly behind the hometown cheering section. Someone has set out a three-foot-wide plastic tub of old-fashioned rice wine, called *makkoli,* to dip up for free, and lots of jolly, red-faced farmers are helping themselves while waiting for their friends' performance.

The dance is a hurricane of color, blue, red, and yellow racing in front of me. Two dozen dancers beat drums and brass hand gongs as they march complicated figures, sending the eight-foot streamers on their hats whipping in frenzied curves through the air. Some dancers even manage acrobatics on top of this. The driving rhythm is infectious—some of the cheering section are dancing in their seats, and one guy who has been helping himself liberally to the free *makkoli* decides he's going to join the troupe, and nearly collides with the whole line of them. The third time he's escorted back to his chair, I'm afraid they're about to hog-tie him, but he lets his buddies jolly him into keeping his seat. It strikes me: the music of this farmers' dance is little different from the sounds of the exorcism ceremony this morning.

And all day long, I'm being force-fed tastes and samples of ginseng, in every conceivable form. Black ginseng extract so bitter I can still taste the drop on the end of the toothpick twenty minutes later. Ginseng wine. Deep-fried ginseng roots, honeyed ginseng slices, ginseng tea (both hot and cold), ginseng chocolate, jelly, hard candy, smoothies, stew. . . . I manage to avoid the ginseng eggs (laid by a chicken gorged on ginseng), and the ginseng honey.

By the end of the day, I've eaten—whatever the actual amount, it's definitely far too much ginseng. My stomach is churning, and when I close my eyes, bright-colored patterns spangle the backs of my eyelids. I wonder how I'll ever be able to sleep. On my way back to the bus terminal, I sit down on a gritty curb on the main street to calm my stomach, and I watch the stream of Koreans headed home with their shopping bundles tied up in traditional carrying-cloths called *pojagi.* It's the week before Chuseok, the autumn harvest holiday, much like a combination of western Christmas, Thanksgiving, and All Soul's Day, and they're laden with presents they've bought for their families: huge gift sets of ginseng products, ornately boxed and gaudily wrapped. With their brand new Samsung camera phones, they grab photos of themselves posed in front of stalls of ancient medicinal roots.

In a strip mall in Wausau, Wisconsin, Diane Zimmerman is having a very bad day at the office. The phone at the Ginseng Board of Wisconsin won't stop ringing, the fax machine is going nuts, and her boss keeps handing her still more numbers to call.

All this is making it fairly difficult for me to carry on a conversation with that boss—a thirty-something ginseng farmer named Joe Heil, the Board's newly elected president. He keeps jumping up to take the phone, the fax machine has a deafening beep, and Diane has urgent questions.

It all has to do with a pesticide.

"We just got a crisis exemption on a new chemical," Heil explains, "because of the weather we've had, and the problems last year—the Environmental Protection Agency calls it a crisis exemption. It's for a product called Gavel that they use on potatoes, and it also helps against phytophthora. There's a whole bunch of hoops you've gotta jump through, to prove to the EPA that it's not only safe, it's also viable, and cost effective, and if we don't get it how it could hurt the industry . . . so that actually we just got it now. So that's why the phone's kinda busy. We've got to call all the growers and let them know there's a new chemical they can use, and the rates."

Even lifelong Wisconsinites are startled to learn that huge quantities of ginseng are farmed there in Marathon County, Wisconsin. Ninety percent of the ginseng produced in the United States grows within twenty miles of that strip mall.

It's an area that's known mostly for potato fields and paper mills, and its rabid Green Bay Packers fans. But every year, scores of Chinese wholesalers make the trek up to Wausau to buy tons of newly harvested ginseng. One of the GBW's functions is to act as a clearinghouse. The office has shelves of white plastic bins filled with roots from different growers. The Board can't help with price negotiations, but it does make possible "one-stop shopping" for the busy herbal medicine dealer. During one particularly frantic outburst of telecommunication, I pass the time by running my hands through the gnarled roots in the two dozen bins. It's amazing how much the lots vary in color, shape, and the anatomy of their limbs and bellies.

The Ginseng Board of Wisconsin was established by the state in 1987. Every time cultivated ginseng root is sold, a certificate is filled out by the grower and buyer and filed with the state, and the grower is billed an assessment. The seven-member Ginseng Board sets the rate, currently 15 cents per pound. This revenue funds the office and supports ginseng research, both agricultural and medical.

Some of this ginseng will be directly exported as whole roots, but increasingly it's processed into value-added products in the United States. It's no surprise to see capsules, teas, and tonics displayed on the shelves in the office. But there are also soup base, beer, boxes of candy, bottles of Arizona Ice Tea—even cranberry-flavored "ginseng chew," promoted as a substitute for chewing tobacco and packed in the same flat tins.

Wisconsin's ginseng harvest reaches up to 2 million pounds in a good year. Unfortunately, the last good year was a long, long time ago.

"We're probably between 500 and 750,000 pounds out of Wisconsin," Joe tells me, gloomily. "Ten years ago, it was in excess of 2 million pounds. We've really fallen, mainly because the cost of production has increased, and the price has decreased, and the weather's been terrible. We've had a lot of rain, three, four years of that, extremely wet, which of course hampers our crop."

Despite the decline in production, prices remain dismally low. "We try to hold, but the buyers come to the farm and they continually tell us they can't pay more money, and due to the fact that everybody's in a financial crunch, we've been forced to sell. Last fall, we probably had a ten- to fifteen-dollar swing in price. Fresh roots were anywhere from $18 to $38. At some points we're actually selling at less than the cost of production. But prices at the other end, in the Chinatowns, and in China, haven't really fallen, so we know that somebody in the middle's making a lot of money at our expense."

Many ginseng growers in Wisconsin have been forced out of the business. At its peak, the Ginseng Board represented fifteen hundred farmers; now there are fewer than three hundred active growers. Because it takes a minimum of three years until harvest, another few hundred farmers still have roots in the ground, but they will soon be digging their final crop.

As in other types of farming, the survivors are the ones who resisted the temptation to borrow money to expand during the boom years. "Ten years ago," Joe explains, "the cost to get into ginseng was probably four to five times greater than it is now. Everything was at a demand. Now that so many people have gone out of the business, it's easy to get in or expand. Ten years ago, seed was $100 to $125 a pound. Last year you could have bought it for $10."

The door opens, and Joe's mood brightens when he sees who's there. "This gentleman here is Kirk Baumann—he's our treasurer."

"Hi! My hands are dirty . . . ," Kirk says, apologetically. He's tall, blond, and energetic, with a classic Wisconsin accent that thumps and polkas its way through every sentence.

"He was out in his ginseng garden," says Joe. (In Wisconsin, it's still a "garden," not a field.)

"We got it?"

"Yep!"

"Cool!"

"And Diane's been on the phone letting everyone know."

"Bill and Bob are in Antigo, right?" she breaks in. "What area code is that?"

Phytophthora, I learn, is a fungus that rots plant roots,

similar to the blight that caused the great Potato Famine across Ireland in the 1840s. When it hits a field, it can spread rapidly, discoloring the ginseng roots, turning them rubbery, and destroying their value. Measures like clearing a perimeter of healthy plants from around the field and disinfecting tools and footwear have limited effects, and the fungus can build up a resistance to pesticides. This is why the growers are so glad to have a new chemical in their arsenal—for now.

"We can use six applications, but that's just this year. Next year, we won't be able to use it again unless we do the same thing or unless it gets registered."

"It has to go through field trials," Kirk ticks off, "residue testing . . ."

"A big concern here in Wisconsin is chemical residue," Joe explains. "We know other places don't care about it, and that is one thing we're promoting in Wisconsin ginseng: our chemical residue is minimal, so it's better for your health. If you're taking it for your health, you shouldn't be taking a product that has a high chemical residue in it, it kind of defeats the purpose. And we use a lot less pesticides and chemicals than the potatoes, hands down. People always talk about ginseng growers—'Those guys use so much chemicals'—but you look at what's used on potatoes, in comparison to ginseng, there's no comparison. It's phenomenal."

"Just a ton of chemicals," Kirk concurs. "A grocery list of stuff."

"And how many potatoes a year do you eat, in comparison to ginseng?" Joe goes on. "Not that it's bad for you . . . nowadays, any chemical that gets registered is so safe."

Diane checks back in: "I called all the big places," she reports. "Marathon Feeds, O and H, the Co-op . . ." She's cut off by a ringing telephone.

Disease problems have changed the whole rhythm of ginseng growing in Wisconsin. Unlike other crops, which are vulnerable only to one growing season's hazards of pests and weather, ginseng takes a minimum of three years to mature, and farmers are increasingly nervous about the gamble they are taking. Traditionally, until around ten years ago, ginseng was always grown for four years before it was harvested. Now, up to 95 percent of ginseng is dug up after three years, by farmers afraid to gamble everything on another year's growth.

"If you've got good looking threes," says Kirk, "some people might leave it an extra year, but it's gotten to be a three year cycle . . . because you never know what the next year brings. And there are a few people out there that do grow some fives—a very few people. Probably can count the acres on one hand."

But all in all, both men are optimistic for the future. If the weather holds decent through fall, if they can get permits for new chemicals next year, if they can get productivity up to where it always used to be, if growers can somehow band together and stand up to the Chinese buyers and demand a fair price. . . . If.

Do they ever worry about the future demand, once China becomes more and more Westernized? Will the Chinese still want dusty, ancient roots instead of the latest pharmaceuticals?

Joe shakes his head. "We hear a lot of that, that the younger generation of Chinese will not mess around with old wives' tales like ginseng, but the demand now is greater than ever before. A few years ago, when Canada was producing their bumper crop of 6 million pounds, and Wisconsin was pushing out 2 million, and then China grew another 500,000, it all got used up. And if you wanted to fill a shipping container to go to China of Wisconsin root right now, you couldn't do it. There's none around. A few little lots around here and there, maybe 40 or 50,000 pounds, but it's probably one of the lowest amounts of inventory Wisconsin has had in years."

Kirk agrees. "The Chinese economy is picking up steam, they're lining up with more money in their pockets, they have more opportunities to buy ginseng and use it." It's interesting to hear farmers in central Wisconsin analyzing economic development in China. "As their economy grows, and they have a lot more wealth over there, they're able to buy a lot more things. It's a tradition, ginseng, over there."

And there's also domestic demand to think about, they remind me. Americans are getting more and more interested in natural healing. The Ginseng Board has been funding several small research projects at nearby medical schools on possible uses of ginseng in treating cancer and diabetes. Both men seem perversely pleased at the skyrocketing incidence of diabetes in the United States—there are indications that American ginseng may be effective in helping some diabetics regulate their blood sugar levels. Any definitive breakthroughs would send the price soaring.

"Do you guys take ginseng?" I ask.

"Oh yeah!" says Kirk.

"Every day," says Joe.

"You might not notice it as a big rush, but if you take it for a month, and then stop, you'll notice it more. It kind of builds your body up, it's good for stamina."

"I always take ginseng in the morning. If I take it at five, six o'clock, I can't get to sleep."

Joe has to go off to get some hay bales delivered, but Kirk offers to take me out to his farm and show me around. I follow behind his well-used truck down a series of country roads, past (yes) potato fields and a paper mill. Then I start seeing dark shade cloth, acre after acre of it.

Around ninety acres, he tells me, when I climb into his truck—his is one of the biggest ginseng farms in Marathon County, and therefore in the entire country. He works it with his father, and hires crews of Hmong laborers for seasonal tasks such as weeding and harvest. "When we first started, back in 1978, I think there were twelve of us, and it took all weekend to plant an acre and a quarter. Now three of us can plant four acres in a day. That shows how much more mechanized we are." In Wisconsin, the sheds that ginseng grows in are so high, and the rows so wide, that tractors can be driven right through them.

As Kirk takes me around his farm, he walks me through the process by which ginseng is raised in Marathon County.

"Ginseng seed," he explains as we stand outside a barn, "takes two years to start growing. The berries are harvested in fall, packed into gunny sacks, and stepped on once a day for two weeks, and we flip them every day to prevent decomposition. Then the berries are hosed down in a tub to wash off the rotted pulp. After that, the seed is mixed with sand, and buried in boxes." Standing atop a twenty-foot-long mound of sand, Kirk gently digs down and unearths the corner of a heavy metal box. Next year, he'll sift the seeds back out of the sand for planting.

Back in the truck, we head out toward his ginseng fields, past stands of the other farmers' more prosaic corn. "We plant anywhere from the middle of July, right up till when the ground freezes, probably until November, even. The way that ginseng is grown here is in beds that are approximately five foot wide, with a foot for the gutter where the tractors drive."

He stops the truck, and I step down into the brown, sandy soil under the black cloth canopy. "These are some threes," he says. "We'll start harvesting them anywhere from the second week of September until early October . . . these are looking pretty good." I look out over the rows and rows of ginseng plants, growing on long mounds with slight depressions between them to channel the water away. He hunkers down and digs around one root with what looks to be a screwdriver, pulls the plant up and nods, satisfied. "Typically you're harvesting from October until November. Although if you're having problems with a garden or rot is moving in, you might harvest in the middle of summer, before more shang rots away. Shang—that's the slang for ginseng."

Inside a shed, he shows me the potato digging machine that's used for harvesting. "This is a barred chain that vibrates, as a knife goes through," he says, pointing to a broad belt of metal links. "Dirt and roots go onto this chain, and the dirt falls through and the ginseng stays on top of this chain, and falls on the ground in back of the digger. People pick up the root and put them on wagons—we pick them into a five gallon pail, and people toss that onto the wagon. Then it's taken to a pond—you're by a pond, always—where the ginseng is washed, all the dirt is taken off, and it's put on sections of some type of a racking system."

Here's another example of ingenuity: some farmers have adapted tobacco kilns to dry ginseng roots. Inside a barn, Kirk shows me the system he has installed, with wooden trays about four feet by eight feet that stack tightly into a chamber. The roots are spread out in the trays, and a furnace forces 100-degree air through them, with a fan to circulate the air quickly. Other fans draw the humid air out of the building. This process goes on for about two weeks to get all the moisture out of the ginseng. Then the roots are packed into fifty-gallon silicone-lined paper drums to prevent the ginseng from reabsorbing moisture. Each barrel contains about one hundred pounds of ginseng.

In a corner of Kirk's barn, dozens of these barrels are lined up, stacked, waiting for the Chinese buyers to come and sample it. It's all from last season's harvest, but he's still crossing his fingers, and hoping for a better price.

As he follows me back to my car to say goodbye, his big, amiable dog tags along, just to check out what's happening. The afternoon has turned gray and threatening, and I can see I'm going to get stormed on, just down the highway.

"How," I'd asked Joe Heil, "did Wausau, Wisconsin, ever get to be the ginseng capital of America?"

"Well, it started with some people called the Fromm Brothers," he told me. "They got seed from someone—they traded pelts for ginseng seed or some goofy thing, and they were actually large growers. The price went to hell, and they went bankrupt."

Later, much later, I learn that that's a pretty accurate summary, except for one key detail. Heil got it exactly backward: it was the ginseng that led to the pelts. The Fromm Brothers of Hamburg, Wisconsin, pioneers of large-scale cultivation of ginseng and founders of a lucrative industry, were just trying to raise the cash to start a fox farm.

I unearthed their saga in a curious book published in 1953, catalogued under "ginseng" in a big university library, but entitled *Bright with Silver*—a color that ginseng most definitely is not. Written by a woman named Kathrene Pinkerton, it reads something like a breathless cross between corporate history and old-fashioned children's adventure novel, punctuated with plenty of moral lessons and exclamation points.

The Fromms were a large and prosperous family who lived on a farm twenty-two miles outside of Wausau. Sometime around the turn of the twentieth century, Henry Fromm, the youngest of the boys (there were six of them) got the notion that a pure silver fox must be the most beautiful animal in the world, and that he was going to raise them. According to the book, he was seven years old, and all he knew of silver foxes came from a magazine article his older brother had read aloud to him about a single pelt that had been sold in London for $1,200—over $20,000 in today's currency.

That was enough to get the Fromm boys going. Silver foxes were a genetic variation of the common red fox, born from time to time in the far north, and exceedingly rare and valuable. Reports of them occasionally turned up in newspapers. In 1901, of the 7,305 fox pelts sold by the Hudson's Bay Company, only 325 were silver foxes—and some of those were almost black, with just a few silver hairs. Henry and his brothers roamed the woods hunting and trapping, and sold pelts to earn pocket money. They had intimate knowledge of woodland animals and their ways, and thought that if they could just get their hands on a pair of silver foxes, they ought to be able to breed them and raise litters of silver cubs.

There was just one problem: money. After endless attempts to trap a silver, they finally decided they were going to have to buy their first breeding pair, and the price was more money than a farm boy could dream of saving up in a lifetime. A live pair of silver fox pups would cost them $35,000—if there were any available. Their father, sensibly, thought the idea was absurd and refused to help. Then the boys heard about a neighbor who was trying to grow a weird new crop—a strange root that Chinese people used for medicine. He claimed it was worth $20,000 an acre.

Wisconsin fur trappers had long collected medicinal plants like ginseng to bring in a little extra cash, but they'd never actually tried to cultivate them. The Fromm boys planted their first ginseng field by resorting to good old-fashioned thievery. Prowling around neighboring farms, they dug up wild ginseng plants wherever they found them in the woods and transplanted them into a bed they had prepared. Their neighbor, reasoning that forest plants would need to be shaded from the sun, had built an arbor over his, and the Fromms improved on this by making sections of screen out of wooden lath. If their first experiment didn't work out, they figured, they could take down the sections and use them over again.

The boys started out in 1904 with 150 stolen wild plants, which were not easy to assemble. Many of the ones they dug up were too old to grow well. Every year, when the plant dies back for the winter, the stem falls off and forms a scar on the top of the root. You can tell the age of a ginseng plant by counting the scars. On some of the plants they poached, the Fromm boys counted forty or fifty scars. One plant was seventy-five years old. The brothers soon realized they'd never be able to find enough viable plants, so they began to experiment with planting the seeds as well—and made the discovery that these needed eighteen months just to germinate.

It didn't take long for other farmers to hear about the new crop, and soon there were no wild plants left in the woods. It also didn't take long for the poaching to start. When one neighbor had his entire planting of ginseng stolen, the Fromm boys rigged a burglar alarm system with trip wires that would ring bells and flash lights when brushed by a thief (or a squirrel). Finally, four of the boys took to sleeping on top of the lath screens that shaded their ginseng garden.

Though it would be years before they could harvest their first crop, the boys kept planting more ginseng, and calculating and recalculating just how much it would take to buy their pair of silver foxes. It meant a tremendous amount of labor. Humus had to be scraped off of nearly every inch of the farm to provide the deep soil the ginseng plants needed. To plant a half-acre, they had to gather and process seven pounds of seed (at eight thousand seeds per pound), plant it in seedbeds, and then transplant thousands of plants by hand. The boys calculated that five years later this would bring them $10,000—if the price held. Already there was much concern over whether the Chinese would still want their roots when they were ready for market years later.

The Fromm boys kept tinkering and experimenting. They planted larger and larger fields. In 1912, when they harvested the first field planted from seed, they rigged up the motor of the family car to pump water for washing the roots—a miserable task in raw November in Wisconsin. Every year, their ginseng harvest increased, and they plowed all the proceeds into buying promising-sounding half-breed foxes to cross, with little to show for it except one disaster after another. One pair, allegedly "mixed blood," turned out to have not a drop of silver fox blood in them. When they bought another pair at great expense, the male killed and ate the female. Prices of silver pelts were getting even more fantastical, but after ten years of planning and five years of breeding foxes, they still hadn't produced a single silver pup.

Meanwhile, their ginseng business was taking off. By 1914, they had to hire teenagers from the neighborhood to help with scraping soil and planting. Harvesting their first large crop in 1915 gave them the confidence to persuade their mother to mortgage the family farm, which was in her name, to buy more breeding stock in their quest for the silver fox. Their first big half-acre field brought them far less than they had hoped, though—five years of backbreaking labor earned them $3,564.09.

Then the price of ginseng started to slide—from $7 to $5 a pound for top-grade roots—and they ran into an insoluble puzzle. Once ginseng had been cultivated on a particular piece of farmland, it could never be grown there again. The seeds would sprout, the plants would begin to grow, then they would all become deformed. To this day, the cause of the "once and done" phenomenon is unknown—Joe Heil theorizes that it has something to do with diseases in the soil. But the Fromms had a ready solution: buy more land. Their 1919 harvest sold for $40,000, and they bought a neighboring farm and kept on expanding.

Their fox breeding efforts had finally produced a single silver pup in 1913, and though the genetics was very poorly understood, they gradually managed to produce a few more—then stubbornly decided to aim for lighter and lighter silver fur, rather than the darker pelts that were the height of fashion then.

But in their ginseng production, even they bowed to a greater authority, recognizing that the Chinese nation was unlikely to change its tastes in ginseng. They experimented with ways to produce the types the market wanted: for example, not washing it very thoroughly because the Chinese liked darker roots. Throughout the 1920s, the Fromms were producing enormous harvests, worth up to $115,000 annually, and investing all the profits in fox cages, fox feed, and fox research. Their yield of two thousand pounds of ginseng root per acre is something Joe Heil speaks of wistfully.

The Great Depression didn't hit Marathon County until 1931, but when it did, it was devastating. Not only did Americans lose all interest in fur coats, but war was on the horizon, following the Japanese invasion of China, and the ginseng market dried up overnight. If carefully processed, the roots could be stored for long periods in hopes of a better market, but running their ginseng farms took over $40,000 a year in operating costs. After much discussion, the brothers decided to keep on growing ginseng and store it all in hopes of an improving market.

They stored their 1931 harvest, and their 1932 harvest. They did it again in 1933. It was not until World War II ended that they were finally able to sell it—by which time they had fourteen years' worth of ginseng harvests packed away in barrels. Ginseng is not a crop for the impatient. Finally, in 1946, they were able to ship their roots to China—and the price they got made the whole gamble worthwhile.

The Fromms picked up right where they left off, tinkering and mechanizing and modernizing. By 1951, they were planting eighty acres of ginseng a year once again, from a low of ten acres during the Depression. They were experimenting with new pesticides, some developed in their own laboratories. "The Company dusts from mid-May until mid-August, and uses 75,000 pounds of chemicals yearly," the book reports proudly. There was still no solution to the problem of disinfecting soil, so the Fromms just kept on buying more farms.

At the time *Bright with Silver* was written, in 1953, their operation was virtually an empire. The 160-acre family farm, once mortgaged to buy three foxes, had grown to 17,000 acres. There were boarding houses for employees, laboratories, acre upon acre of ginseng, and pens for seventy thousand animals. And here the book triumphantly concludes, confident that the dream of the Fromms will continue to unfold, chapter after chapter, that the seemingly unlikely combination of silver foxes and ginseng will spin prosperity forever into the future. In the closing words of her book, Kathrene Pinkerton says the company "holds a wealth of experience and knowledge in the pioneering of this continent's last natural resources, and whether the fashion is for long- or short-haired fur, or no fur at all, there will always be 'The Company.'"

Except now they're not in the phone book, they're not in any database, and at the Ginseng Board of Wisconsin no one has any idea what ever happened to them.

I finally learn their fate when a helpful librarian at the Marathon County Historical Society unearths and mails me a copy of a melancholy newspaper article from the Merrill, Wisconsin, *Foto News,* dated thirteen years back. The photo depicts high weeds growing through sagging frames of wood and wire mesh. The caption: "Broken and run-down fox pens are the only sign of a famous and world renowned fur business in the Hamburg area."

The reporter spoke with Ned Tead, a descendant of one of the brothers, who in 1986 became sole owner of the business—and its $2 million debt. "There was no research and no innovation to keep the business alive," he told the reporter. "The history of the company is painful." He admitted that the brothers were notorious for paying starvation wages, even to relatives like himself, and called them "uneducated farmers," prone to making senseless business decisions. Lawsuits split the family apart, and everyone suffered financially.

To pay off the debt, Tead sold most of the remaining Fromm land—along with the sixteen farmhouses and the last of the foxes. In 1988, the final ginseng crop was dug. Fromm Brothers morphed into an auction company, but by 2003 that had vanished, too. It's a classic American rags-to-riches story—except it ends again in rags.

Months later, a Marathon County ginseng insider, who doesn't want his name used, gives me his perspective on the Fromm saga. Sure, I knew them, he says. I knew the family and the people who worked for them. The grandson came complaining to me many, many times—"It's such a hard way to make a living!" I was making good money then, but that grandson kept saying he couldn't make a living.

Well, the man tells me, it depends on what kind of living you're talking about. Every six months that grandson would have a new sports car. And he'd just go cruising around, and not tend to the business. By the 1970s they closed their office in Hong Kong, and in 1987 they auctioned everything off. He committed suicide six months ago. He was forty-something years old. Very sad, the insider agrees.

But then he tells me what he figures is the real reason. "The Chinese saying is, *Riches won't go past the third generation.* And that," he says, "was exactly the third generation."

FOUR

VIRTUALLY WILD

In a chilly meeting room in Rockcastle County, Kentucky, on a Saturday morning in March, fifteen people are leaning forward in their chairs, looking to get rich. They've driven in at the crack of dawn from around the state: from Harlan County, deep in the mountains; from Frankfort, the state capital; out west in Hodgenville, where Abe Lincoln was born; Laurel County, just down the road.

The workshop they're attending is sponsored by an organization called the Appalachian Ginseng Foundation, but the ginseng-growing technique it teaches is the brainchild of one man—a Kentuckian named Syl Yunker, who has been developing and promoting the ginseng he calls "virtually wild" for fifteen years now. His theory is seductively simple. By planting ginseng seeds in the plant's natural environment in undisturbed forest floor, and just letting nature take its course, a landowner can, with virtually no work at all, produce roots indistinguishable from the wild variety—and worth a not-so-small fortune. Some say $70,000 an acre, every year.

An intense man with steel-gray hair and a neat white goatee, Yunker is quietly articulate, with a missionary's obsessive zeal. In the past five years, he has given workshops like this one across

---

Previous page: Ginseng grower Syl Yunker in his forest, displaying a tool he has invented for planting the seeds.

Virginia, West Virginia, Tennessee, Kentucky, and North Carolina, the Central Appalachian heart of ginseng's natural range. "The future of virtually wild ginseng here is almost unlimited," he proclaims to anyone who will listen, including this audience in Rockcastle County. (In his soft Kentucky accent, the plant's name comes out sounding like "Jenson.") "There will be no other location in the world capable of growing it for at least a thousand years. Say China gets into reforestation. . . . They can grow a forest in seventy-five years, but they can't reproduce the soil. That would take a thousand years or more."

But these people haven't driven all this way for the grand sweep of history. What they want are the nitty-gritty practical details on how they can grow ginseng on their woodlot—and that's exactly what Syl gives them at the workshop, in great abundance. How they should plant, for instance.

"This here's a little device a number of us came up with, to do your seeding without a lot of stooping." He holds it up: it consists of a thin wooden pole four feet long, a white plastic tube, and a knife-blade, all bound together with liberal amounts of duct tape.

"With one of these Kroger-store mop handles," Syl instructs his listeners, "you take a good, strong knife—this here's a Bowie knife—and you tape it on there so that you've got at least four inches beyond the end of the handle. The idea is you're testing for depth. You like to have at least two inches of soil before you put a seed in, and you can jab this in and test the depth of your soil. The mop handle protrudes a little beyond the end of the plastic pipe where the seed's gonna come out, and that prevents the soil from getting up there into the pipe and clogging it up—that's just a half-inch diameter PVC pipe. . . . So you jab it into the ground, and when you're sure you've got a proper depth, you wag it back and forth, with the blade sideways. . . ." He demonstrates so energetically it looks like he's about to bore into the floor. "Then," he continues, tilting the device at a diagonal, "you kind of lay it so the seed will slide right off the edge of the blade." He aims an imaginary seed with thumb and forefinger, drops it into the pipe, and watches, satisfied, as it lands in the imaginary hole. "Then kick your leaf litter back over it."

"Syl, you could use that thing on poachers too, couldn't you," calls a voice from the audience.

Syl eyes the blade, gravely. "It's a good, stable walking stick. When you get on too steep a slope, you can jab that knife in. . . . I've used it for a number of things," he allows. "I was glad to have it with me when I ran into a black bear, once. I think she was in labor, having her young."

"You didn't pester her too much, did you, Syl?" a voice calls.

"No, I just *eased* away . . . *eased* away. . . ."

The audience is in stitches, but Syl goes on, deadpan. "I could hardly believe it, because I'd never experienced anything like that before. So I stayed around long enough to find out. My attention was drawn to a very heavy breathing." He imitates it: a slow, rough gasping. "And I thought, *uh*-oh! That sounded awful near. . . . So I stopped, and just kind of froze on the spot—am I hearing what I'm hearing? I looked back over my shoulder and there was a great big uprooted pine tree, with a deep hole underneath it and brush and vines grown over. I couldn't actually see her. I convinced myself, well, I'm gonna walk away slow and count my paces, see how long I can hear. I was fifteen paces away before I couldn't hear her."

Afterward, his suspicions were confirmed. Later that year he encountered a black bear cub near the same spot. "I talked to the Forest Service later, the animal people, and they said yeah, that's about the widest swing they make going from the Carolinas up this way to West Virginia, when they migrate." All of this took place only ten miles from a town with traffic lights and strip malls.

Yunker speaks slowly and deliberately, in a soft, husky voice that gives every word its own weight. Some of the people that come to his workshops, he acknowledges, are from families that were 'sanging before he was born. "I don't make any claims of being an age-old expert," he says. "I got introduced to ginseng when I was eighteen, in the service in Korea." It's only when he mentions that this was in the 1940s that I stop to subtract and realize that he's a couple decades older than I had figured. (And yes, he says, he consumes plenty of ginseng.) His brother-in-law, Bill Bass, had worked in China for British-American Tobacco before World War II, and stayed on postwar after the U.S. military took control of Korea, to work with the Korean government agency administering the national tobacco and ginseng monopoly.

For years after his time in Korea, Yunker was gripped by ginseng's possibilities. In the early 1980s, he began trying out a new idea on some land he then owned near Bybee, Kentucky, seeding ginseng in the forest floor and then letting it grow undisturbed. "It's been one big, long experiment," he says. "Because the whole concept was a self-renewing crop, the first question was, what is the carrying capacity of the piece of land?"

He investigated how many seeds to sow, how far apart, how often. He looked at different types of soils, how deep the shade should be, and how high on a slope to plant. As signals of good locations, he noted the kinds of plants that most often grew near wild ginseng—things like bloodroot, black cohosh, and goldenseal—and the trees it seemed to have an affinity for. Maple and poplar, he noted, were two of the best.

Later, he located and bought what he was pretty sure would be a perfect patch of Kentucky ginseng land, further to the east. The experiments went on. He planted on different degrees of slope, seeded at different distances, began forming terraces with the trunks of dead trees to keep the autumn leaves on the hillside. He confronted problems not commonly faced by most farmers, like wild turkeys devouring his seeds.

And there was another factor he hadn't counted on—the poachers. In a small mountain town, the roots can easily be turned into quick cash at the local flea market or grocery store.

"Oh yeah, I've been poached," Syl tells me. "I caught them in the act. I identified them, and took everything off them including their clothes, and searched them, and I walked them back up to their car. Then I went down to the sheriff, and said, OK, what can we do here? And he said, all we can establish is that they trespassed. Well, I said, they had over $100 in ginseng, and that in Kentucky is a felony. And he said yeah, but they're going to say they got most of it somewhere else. And if you charge them, it will come out in the local paper, and you'll have the whole county up here."

The sheriff finally offered to "go talk to them." Did it work? "Well, evidently he did a good job, because I haven't seen anybody up there since."

To the audience creaking on their metal chairs here at the workshop, Syl's method needs no sales pitch. Its advantages are clear. For one thing, it's perfectly suited to people like them who have land but little cash, who are hoping to grow ginseng without a huge up-front investment. Growing in the forest avoids the labor and expense of setting up artificial shade.

Though growers in Korea and North America have developed forms of artificial canopy that produce a good approximation of natural forest shade, they are extremely costly. For an acre's worth of canopy made from dark polypropylene shade cloth, a grower must count on investing at least $10,000. That acre of cloth weighs nearly seventeen hundred pounds, and takes considerable expertise to set up. Posts have to be sunk and cables strung, and after that, three workers will need an entire day to install the cloth, assuming they already have plenty of experience tightening large cables. For the small farmer, it can be an overwhelming project, and the whole system will have to be replaced after twelve years.

Some growers feel that shade structures built of wooden lath produce better roots, but these are just as expensive, and even more labor-intensive. In his book *American Ginseng: Green Gold,* Scott Persons provides detailed plans and helpful instructions for constructing a four-foot by twelve-foot wood-lath shade screen for ginseng—and then gently informs his readers that they will need to build and erect a thousand screens like this to shade a single acre. And these, too, will need replacement after twelve years.

Furthermore, virtually wild ginseng doesn't require fertilizer. The fallen leaves from the hardwood forest it grows in provide all the necessary nutrients. Winter snowfalls compact the leaves and cause them to decay where they fall, though on steeper slopes growers may need to form terraces with dead branches to keep the leaves from slipping downhill. "With cultivated or woods-grown, you'd be paying for all this fertilizer," Yunker points out. "And it wouldn't have all the nutrients."

The virtually wild technique also sidesteps another problem that plagues growers: disease. As ginseng activist Ann Rogers of Roanoke, Virginia, explained to me, "Ginseng is like any kind of monoculture crop, like cattle, for instance. You pack them too close together and you're going to get disease." Wild ginseng grows naturally in patches, separated by numerous other types of plants on the forest floor, so pests do not spread easily from one location to another.

Intensively cultivated ginseng, even when grown in the forest, is subject to a host of ailments, simply because the plants are in such close proximity. The chemicals used to combat these problems sound alarming to the uninitiated. For example, to prepare the soil for planting, some experts recommend fumigating it—applying preparations that will kill every living thing in the soil, including seeds, spores, roots, insects, and worms. A typical fumigant is formaldehyde.

After the ginseng is growing, it provides a feast for all manners of creatures, from molds to slugs to chipmunks. Again, for most intensive growers, the usual strategy is chemical. One ginseng primer lists twenty-two different pesticides that are effective (and labeled) for use on ginseng. Another book describes a potent chemical concoction that is actually illegal for use on ginseng—unless users mix it themselves. Besides being costly and dangerous, pesticide use raises concerns about its effects on ginseng consumers. And ironically, these are exactly the people who are most concerned about their health and natural approaches to healing.

Because virtually wild ginseng plants are spread out over a large area, and interspersed with many other naturally occurring plant varieties, diseases have a much harder time taking hold. The most common problem, alternaria blight, is carried by spores on the wind, but seldom spreads far through virtually wild plantings.

The whole key to success with the virtually wild method, Syl emphasizes over and over to his listeners, is picking the right spot—the right land, and the optimal location within it. Once the seeds are in the ground, the grower does nothing, literally, for ten years. How well the ginseng grows, and its ultimate value, depends entirely on the location that was chosen. The amount of light the plants receive is regulated not by artificial canopies, but by the direction of the sun and how much filters through the trees to the forest floor. The nutrients that fuel growth come not from big sacks, but from the built-up soil and the autumn leaves drifting to the ground. And every foot of land is different.

Syl picks up a marking pen, and on a big, white sheet of flip-chart paper he gravely writes "SANG," his mnemonic for the key factors in choosing where to plant. The S stands for several things—first is "slope." Ginseng grows best on hillsides, where the drainage is good, and the most favorable slope is 40 percent. Any more than that, and the grower will need to make terraces to keep the autumn leaves from heading straight downhill, leaving the ground bare. It's also easier to work there, when the drop isn't more than 12 degrees—"You don't have to have one leg that's *too* much shorter than the other." And the best orientation for the slope is between north and east. "In the summertime, you get a little bit of morning sun, but by noon the sun is behind the ridge and you get a cool afternoon and evening."

Another S is shade—about 70 to 80 percent is ideal. Cultivators spend thousands of dollars and hundreds of hours of labor trying to reproduce this effect, the cool, shifting patches of dappled light at the floor of the forest. The virtually wild grower merely needs to plant in the right spot. "You can look to find where the snow is staying longest in spring," Syl tells the audience, "and that will be your good area. You're not going to get the sun hitting it in the hottest part of the day."

His third S is soil fertility, which is linked to the types of trees growing on the land. "An area with a variety of trees will have good calcium content," Syl advises. "My land's got maple, poplar, hickory, oak . . . some nut trees scattered in there, like black walnut. . . ." The best soil is under those types of trees that drop their leaves right after the first frost, because the nutrients don't have time to retreat from the leaves back into the tree. There follows a long and detailed discussion among audience members as to just what types of trees, and how many, and where, are on their own land. One guy admits to having "some maples and a *nawful* lot of briars" on his most promising piece of ground.

Syl moves on to A, after hunting all over for his felt pen and at last finding it sticking out of his pocket. Air, in the sense of air movement, is the next important factor. Many ailments that ginseng is prone to are spread by airborne spores. In cultivated ginseng, disease is kept at bay by applications of pesticides. Virtually wild growers depend instead on a good breeze to keep them away.

N is for nutrients (two felt pens have now vanished), and here the discussion turns technical. It's obvious who's already farming, and who isn't, from the questions that are asked. Syl explains that the two main factors needed are good calcium content and the correct soil pH, between 5 and 6. "You get too sweet a ground, and you'll have problems with diseases," Syl advises. In order to grow ginseng that is truly indistinguishable from the wild form, growers must pass up many of the remedies that are allowed even by organic farmers. "Gypsum is the only approved additive, as far as I'm concerned. Or eggshells are OK."

By the time we get to G, all felt pens have headed off to parts unknown, and Syl has to drive home his key points without them. "Remember your *grade.* If it's above 12 degrees, you need to terrace your deadfall." Of course, for some of the audience, all these reminders are academic. Ginseng has been in their family for decades, centuries. They need only wander their land to see where it grows wild, or track down someone who remembers exactly where grandpa's 'sang patch used to be.

After choosing the location comes the only real labor involved—sowing the seeds. A half-acre plot requires four thousand seeds the first year and takes eight to ten hours to plant—not an easy task on sloping terrain, threaded with tree roots and studded with rocks. To reproduce the widely spaced patches that ginseng naturally grows in, Syl advises dividing the field into sections a yard square, and planting one corner per year in each plus a plant in the center, for five successive years, in a checkerboard pattern. The seeds are planted in twos (because each berry contains two seeds) and must be placed a half-inch deep—thus the planting tool with its sharp blade is also used for measuring.

And then, he tells us, just sit back and wait. For ten years. Though cultivated ginseng can be sold in as little as four years, ginseng raised in the forest without fertilizers or pesticides grows much more slowly. The roots must force their way down through heavier, compacted soil. This gives them the characteristic growth rings that distinguish them from the cultivated variety. When they hit a stone or a tree root, they branch, turn, and develop those peculiar limbs that shape them like a human body.

Besides planting more and more ginseng in his woods, Syl put those ten slow years to good use, trying to figure out the best way to get the future crop to market. He started by running an ad in the local paper, inviting anyone growing ginseng, or interested in growing, to attend a meeting. About twenty-five people showed up—some of them, he's convinced, mainly interested in finding out where there was ginseng to be stolen. But the bona fide growers continued meeting, and over the next three years they organized a co-op of seventeen members and set out to study the market.

They started out by mailing a survey to hundreds of ginseng dealers, whose names they obtained from state licensing authorities, asking about their prices and the attributes of higher-quality roots. They got no responses at all. Yunker believes that for dealers, keeping growers in the dark about the true value of their crop is a prime money-making strategy. He charges that by refusing to clarify what they're looking for, they are able to pay far less than the market price for higher quality roots.

Yunker and his neighbors persisted. They promised to share the price information only with people who responded to their survey. Over time, they were able to wheedle responses from 20 percent of the dealers, which were compiled and analyzed by an economist—Yunker's daughter. The result is a "scorecard" of fourteen ranked and weighted factors, including characteristics such as shape, rings, color, weight, taste, and smell, which can be used to rate a single ginseng root or an entire batch.

This information, Syl tells the would-be growers at the workshop, will give them more leverage in negotiating with brokers. By knowing the attributes that the Chinese consider most valuable in a ginseng root, growers will be better able to assess the true value of the roots they have harvested.

"You go to the local dealer and tell him your asking price, and he'll say, 'I know more about this business than you.' OK, fine, you might know more than me, but you don't know more than three hundred people. If you don't like it, you talk to your buddy, and I can talk to him, too.

"I have to use the method of holding their feet to the fire," Syl says, "because they're going to say anything they can to get the price down. The standard answer they give, every year, is, 'That ain't what the Chinese are looking for this year.' Well, the Chinese have been looking for this root here for three hundred years, and over there for five thousand years, and one year doesn't make any difference."

Part of the problem, he feels, is the long chain of middlemen between the grower in Appalachia and the user in Asia. Syl explains, "You've got the food store man who'll buy a little bit, and a few buyers at the flea market—they're putting together twenty, thirty pounds. Five from you, five from somebody else.

And the next guy is putting together fifty to a hundred pounds. He's selling to another person who's putting together three hundred or five hundred pounds. And the exporter ends up with a thousand pounds to a ton before he ships it over. You've gone through four or five layers, each getting a cut—for doing *what?* After you dig it, and they buy that root, it ain't going to change, there's nothing done to it, no value added, from there until it hits Hong Kong. The value added is all in the way you grow it, and the patience you exercise in growing it."

Syl goes on to tell of being approached directly by exporters, one in Portland and the other in Seattle, who offered him the equivalent of $860 per dried pound of roots—at a season when he had none available. Locally, he would have received about $400. "You *can* export it yourself," he says. "They're looking for a volume of twenty pounds before they'll talk to you. I can give you the telephone numbers right now, and you can start calling them." Several listeners nearly bounce from their chairs in their eagerness—clearly, they're not novices after all.

But now it's time for vegetable soup and turkey sandwiches on white bread, and a walk outside in the chill before earliest spring. People cluster in twos and threes, the herbalists asking earnest questions, the family from down the road chatting with the guy from two counties over. And Sister Therese Tackett of the Appalachian Ginseng Foundation hands out plastic bags filled with the burgeoning, beckoning seeds.

"Ginseng," proclaims an AGF fact sheet, "is the silver bullet and key to liberation of the Appalachians from the shackles of the extractive coal, timber, oil, and wood fiber industries." Along with the flow of *chi,* ginseng also seems to stimulate the flow of hyperbole, but in this case there is experience and research to back it up.

"High-grade wild or virtually wild ginseng," their fact sheet continues, "has a proven market that could climb into the billions of dollars as the Asian economy recovers. Ginseng production among central Appalachia's private woodland owners offers a viable option for small-time tobacco farmers who are unable to compete with extensive crops such as kenaf or hemp, or agricultural innovations such as llama or bison raising. . . . The Appalachian Ginseng Foundation has been established to give supplemental employment to woodland owners and small farmers, save our woodlands in Appalachia, and save wild ginseng from extinction."

The AGF argues that ginseng is a win-win solution for Appalachia's economic woes, providing a steady income to small landowners in isolated areas—exactly the people with the highest poverty rates, and the worst unemployment. Once ginseng has been gradually planted, over a period of five years or so, the crop should reseed itself automatically, in exactly the same way a wild ginseng patch would. No labor at all is required between seeding and harvest. Even a small patch of forest can yield a tremendously valuable crop. The market is huge and growing; no other area of the world will be able to produce it anytime in the near future. And, perhaps most significant of all, virtually wild ginseng provides a powerful incentive for preserving the Appalachian forest, because the plant will grow only in undisturbed hardwood forests. Ginseng is infinitely more valuable than a one-time crop of old-growth timber.

The AGF was founded in 1998 by thirty ginseng growers who hoped to spread the word about their innovative business, and headed by director Al Fritsch, since retired. Fritsch is a native Kentuckian, a Jesuit priest with a doctorate in chemistry who cut his activist teeth with Ralph Nader's Center for the Study of Responsive Law in Washington, D.C., in the early 1970s. He has barely taken a day off since. After seven years as director of the Center for Science in the Public Interest in Washington, he returned home to tackle the problems of his native region by founding Appalachia-Science in the Public Interest. Of the numerous projects he has worked on—a weekly television show, environmental resource assessments carried out for two hundred different groups, a score of books on topics ranging from theology to garbage—ginseng seems closest to his heart.

"We come at it from an environmental standpoint," says Fritsch. "We have a mixed mesophytic forest, the oldest and most varied hardwood forest in the world, and it's getting ready to be destroyed. This is a way of saving the forest. It's the most valuable nontimber forest product you can grow. And we want to save our small farmers—my dad was a small farmer." Fritsch grew up on a tobacco farm in Mason County, Kentucky, and once calculated that his youthful labors killed sixty people—smokers who consumed his family's crop.

"And the fact is," he continues, "unlike any other industry we're trying to set up in this country, we don't have to worry about the market. The market is *there.* We just have to be able to get our product to market, and to get a fair price for it. They'll buy as much as we're willing to sell."

Appalachia-Science in the Public Interest, the AGF's parent organization, was founded in 1977 to better people's lives in the region by developing and demonstrating appropriate technology. The dark green building that houses ASPI's office in Mt. Vernon, Kentucky, looks a bit quirky, even by the eccentric architectural standards of a small mountain town. Its parking lot is bordered with burgeoning vegetable gardens in raised beds, and one end of the building's roof is covered with solar panels, which, on a sunny day, actually generate so much power that it feeds back into the electrical network, making the building's electric meter spin backward. (They'll be happy to show you how to set up your house to do the same.) Parked next to the front door is their home-built solar car, a nondescript gray sedan that runs on batteries charged by the solar panels. It's a big hit with visiting school groups. "I get some double-takes when I drive it around town," says staffer Joshua Bills. "When I stop at a light, there's no sound at all from the motor."

The AGF has been busy in the last few years. Besides organizing ginseng workshops in five states, they've produced a lengthy growers' manual. They publish a quarterly newsletter with tips from growers, market updates, and reports on ginseng research. They've written several technical papers on ginseng, including a summary of its role in Appalachia and a numbingly thorough review of potential regulation of wild ginseng as an endangered species.

They've also come up with some ingenious, small-scale techno-fixes for ginseng growers. One of the cleverest is a computerized method for determining optimal planting locations, worked out by a staffer named Dan Bond, whose considerable talents range from banjo-picking to website design. "A farmer knows his land, but there are people who have just come back to their grandfather's farm and want to grow ginseng, but don't have twenty years' farming skills," says Bond. "We're hoping this will be a big help."

Bond's technique combines data from Geographical Information Systems with GPS satellite coordinates to pinpoint the best location for a ginseng field on an individual farm. Growers determine the GPS coordinates of their farm, which Bond overlays on Geographical Information Systems information on elevation, soil types, and water flow. "By combining where the sun comes up with slope data, we can see which areas meet the growing requirements," he explains. "We can add soil data as well."

Bond knew he was on the right track when he tested the system on his own land and found that two of the small areas highlighted were places where his grandfather had once had ginseng patches. Landowners can contact Bond through ASPI and send him the GPS coordinates of their farms, and he'll send back the coordinates of the best locations for planting ginseng, for a nominal charge.

There's just one small problem with this service: it hasn't had a single customer yet. People are reluctant to tell even the Appalachian Ginseng Foundation that there's ginseng on their land, for fear of theft. The AGF has even had to remove its name from the outside page of its newsletter, lest nosy neighbors find out what's growing on subscribers' woodlots. A cooperative they founded deals in "medicinal herbs"—not ginseng.

And these fears are not overblown. As an informal support

group and grapevine for ginseng growers, the AGF hears plenty of sad stories. "One fellow called, almost in tears, because poachers had gotten his whole crop and the sheriff there refused to do anything about it," Al Fritsch told me.

Our Saturday morning workshop is being held at ASPI's other center, a cluster of buildings on a steep hillside above the Rockcastle River, surrounded by some of the prettiest forest imaginable. During the five hours that we sit in the workshop, the quiet is broken only a handful of times by a car passing on the road. The main building of the Rockcastle River Demonstration Center features passive solar heating and a remarkably pleasant composting toilet. (My only previous encounter with one had been horrifying.) The displays of birds and wildflowers inside the nature center distract me from Syl's lecture at times, and it's a good thing our chairs are lined up facing away from the picture window, or I'd be gazing dreamily out at the woods after lunch. Instead I'm wondering whether I might be able to slip in a few seeds on the leafy hillside behind my house.

But a troubling question keeps coming up during the presentation, and though Syl doesn't exactly brush it aside, he doesn't respond to it in great detail, either.

"So," repeats the very young man who's been listening avidly along with his even younger wife, "how *do* you keep the poachers away from your fields?"

This is truly the fly in the ointment, and it's clearly on everyone's mind. When you're raising a crop worth $500 a pound, which will grow only deep in the forest, how do you keep it safe for ten long years?

"Living on the property is your best security method," Syl tells him. Or for absentee owners, he says, it may be possible to set up a sort of share-crop arrangement with someone who will stay on the property and keep an eye on it, for something like a fifty-fifty split of the crop. And here Syl takes off into a meandering digression about his plans to design a solar-heated house that can be built far out in the woods, completely off the grid, with solar photovoltaic electricity and no utility connection but the phone line. He has the ideal dimensions for such a house all calculated, and the guy from down the road starts giving him some suggestions about where to locate the fireplaces, and we never do return to the topic of poaching.

But it's clearly a problem, a big one, and living near your fields won't necessarily solve it. There's a story floating around ASPI of a friend-of-a-friend who had his $300,000 crop dug right out of the ground, on the only weekend in years that he went away. A ginseng buyer up in Ravenna, I'm told, was wiped out in a day. And since many ginseng poachers go around armed, a sensible grower might not necessarily want to have ginseng fields anywhere near the house.

The workshop breaks up in mid-afternoon, and most of the participants are quickly on their way, precious bags of seed cradled on the front seats of their pickup trucks and well-used cars. I stop to chat with Syl and Therese, while the unofficial ASPI dog, a huge mutt of inscrutable ancestry named Freckles, supervises and wanders in the road. "She looked pathetic when we found her," says Therese. "And now she's beautiful! Except her whiskers grow backwards." In fact, they do—they curve forward, like a cat's. Syl is still explaining something to Therese with big, sweeping arm gestures as I drive off through the late afternoon woods, slanting with shadows.

Though Syl Yunker may well be the most colorful advocate for raising ginseng in the woods, he's hardly alone. Overall, the trend in the industry is away from the big, Wisconsin-style plantations and toward smaller scale growing that seeks to emulate nature in varying ways.

To get the bigger picture, I call Scott Persons at his farm in North Carolina. His growers' manual *American Ginseng: Green Gold* has probably touched off more and bigger ginseng dreams than any other piece of writing in history, with its enthusiastic first-person stories and meticulous how-to's. The book has gone through four printings and sold nearly twenty thousand copies since 1986. In it, Persons discusses how to get started in small-scale cultivation both under artificial shade and under natural forest shade. He sowed his first ginseng bed in 1979, at a time when there was virtually no practical information available on small-scale cultivation, apart from decades-old USDA bulletins. *Green Gold* is crammed with insights hard-bought by experience.

But the first thing Persons tells me is that he's now in the final stages of producing an entirely new book. Why tamper with success? I wonder.

"Because the business has changed, and in modern times the rate of change accelerates. For example, the whole business of growing under artificial shade has changed. There used to be a real potential for profit. Around 1994, there started to be an overproduction of field-grown. They began growing this stuff in Canada and in China on a large scale." The price plummeted, leaving only the largest farmers—people like Joe Heil and Kirk Baumann—able to survive. Persons no longer recommends that anyone try to get into that business.

An alternative to the capital-intensive artificial shade approach is woods-cultivated ginseng, in which growers take advantage of natural forest canopy for shade while still tilling the beds, applying chemicals, and trying to speed up the natural growing process. Because the farmed roots grew fast, they lacked the ringed, gnarled appearance of wild roots and fetched a lower price. For a time, woods cultivation produced the most profitable balance between yield and quality. "We lived fairly comfortably, even re-did our house, on the profits from that kind of ginseng," Persons tells me. But over time, the lowered price of farmed ginseng pulled down the value of the woods-cultivated variety. Trying to make their product more distinct, woods-cultivated growers gave their plants more time, letting them mature slowly to get a more wild-looking root.

With some crops harvested after eight or nine years, woods-cultivated ginseng was shading off into a different technique called wild-simulated. Basically, you choose your woodlot, drop in the seed, and walk away. Syl Yunker's "virtually wild" method is one specific wild-simulated technique. Says Persons, "Wild-simulated became relatively speaking more attractive, and as this was happening, the price of wild ginseng was going up, and holding fairly steady. So as a function of that I got more interested in growing wild-simulated, and began doing more of that. The amount of labor required in growing it is far less." It's still a risky endeavor, because in the nine or ten years that the crop takes to mature, a vast number of things can go wrong, from poaching to hurricanes.

A larger portion of his new book is devoted to growing wild-simulated. "Much of the context is around encouraging people in that direction," he says. "Wild-simulated, it mainly requires a good site, and a little experience in how to get the seeds into the ground. It's a much more practical thing to tell people how to do." In 2001, Persons carried out a survey that estimated that about three thousand people in the United States were producing wild-simulated ginseng, and the average area planted had increased over the preceding seven years from a quarter-acre to a half-acre per grower, an indicator that they were having some success. So Syl Yunker is not entirely a prophet crying in the wilderness.

For would-be growers who can't make it to his workshops, Yunker has produced a two-hour video about his method. It's something like a long rambling conversation with a knowledgeable but eccentric uncle. In it, he takes viewers through a year in his ginseng field, from the bare winter, the best time to choose a planting site, through spring, when new growth emerges, on to the summer with its tasks of seed collection and processing, and fall, his time for planting, and harvesting. For the uninitiated, it's a rather puzzling sequence to present things in, but he's been doing this so long it's obviously second nature to him.

"We try in all things to duplicate nature as closely as possible," he says in the film's introduction, and with this in mind, he tramps along through the crackling leaves on the hillsides, pointing out the best and worst spots for planting. High on a ridge, he shows a nine-year-old ginseng plant that's dwarfed by the ballpoint pen he uses as a pointer—"There's too little understory, and it's on sandstone, with not enough calcium in the soil." In another location, lower in a draw with more moisture, he points to a bushy, glossy ginseng plant, and the black cohosh growing nearby, a signal of good land for ginseng. "That's also called rattleweed, and it's sort of a female herb—a substitute for female hormones."

In the realm of instructional videos, his is probably unique in the world. It was certainly the first I'd ever seen that features in its soundtrack the unmistakable sound of large dogs panting. At one point, a shaggy black rump with tail wagging furiously backs into the picture. "Oh, you want to be in the movies, hmm?" says Syl. The video's editing is somewhat quirky, as well. An explanation of the importance of building hillside terraces is interrupted with a nice still from Yunker's trip to China, showing him standing in front of an agricultural museum, and then a correction about when Mao Tse-tung's government took over in China, before returning to Syl on the hillside, firing up his chainsaw to cut some deadfall. Syl's brother-in-law Bill Bass makes a cameo appearance, reminiscing about Korea on his front porch. None of it is scripted, and every bit of it is articulately presented.

But what strikes me the most in watching the video is the unassuming nature of the plant itself. Until Syl's camera zooms in on it, ginseng is just one more little plant on the forest floor, its plain green leaves fanning out flat like a table to catch the flickers of sun, its tiny flowers hardly visible. It's startling to think that fortunes have been made and lost, and people have been shot, over this ratty, bug-eaten weed.

The video's real drama comes in fall. On the forest floor, the leaves of the ginseng and its companion, goldenseal, have turned a glowing yellow, signaling to growers (and thieves) that it's time for digging. Syl is out in his field with a sharp, needle-nose spade, looking for a good specimen. After rejecting several others, he finally settles on a large four-pronger, growing together with a smaller plant a few inches away. "This one's probably ten or eleven years old," he tells the camera. "But you won't know the actual age of the root till it's out of the ground and you count the scars."

He starts by brushing away many inches of fallen leaves, to check the stem as it comes from the ground. Then he carefully digs a circle six inches out, following the angle of the stem, speculating all the while on how the root will look. Two dogs amble over to investigate. When he's dug completely around the big plant, the ball of soil is a foot in diameter, and he tips it free and turns it onto its side, careful not to injure the smaller plant.

The tension builds. Cautiously, he breaks the soil away from the plant, teasing the root free with his fingers. An indescribable shape slowly emerges, like a twisted, crouching animal with many legs, carved from old ivory. "Good size! *Good* size!" he crows, holding it up to the camera. "Here we have our prize." Freshly dug, that single root weighs more than an ounce; correctly dried, it may sell for $20. Then he turns from the camera and gently replants the younger ginseng plant he's dislodged.

After all the tales of poaching and skullduggery I've heard, I'm honored when Syl invites me out to see his ginseng fields. When my husband and I meet up with him at Hardee's in the nearby town, he's drinking a cup of pale, milky tea. We follow behind him in our car, down a state highway, and then down a side road, and finally down an even narrower track. To cover the last stretch, we leave our wimpy, citified sedan and crowd into the cab of Syl's small pickup to jolt along on the rutted track. At the top of a wooded ridge, he kills the motor and we get out.

We're in the western foothills of the mountains, hardly fifteen minutes from town, but it feels like another world. It could be another century. There's no noise of traffic, airplanes, nothing—just bird calls, so many that even Syl doesn't recognize them all.

He spent more than four years just looking for these forty-two acres. It was once part of a family farm, high and isolated.

"There's a coal seam here. The family that lived here before I bought the place, they moved away in the '40s, and they used to get their coal there to heat the place and do their cooking." Aside from their old stone well, still flowing, there's little sign that anyone ever lived here. "They used to grow corn here, but the whole place was all overgrown within ten years."

There's a faint whiff of smoke from a neighbor's distant woodstove, a breath of an older Appalachia that mingles with the damp, fertile smell rising from the newly thawed earth. With the trees still winter-bare, it's easy to trace the features that Syl points out—the long slope from the ridgeline, the creekbed along the base of the draw. "It looks dry," he says, "but the water is there, running underground."

He leads us past mossy limestone outcrops, points out the place where, in wet weather, a fifty-foot waterfall will come plummeting down. And now we're standing on the "bench," the place where he believes ginseng grows best. It's hard to see the extent of his planting, because it's so early in the year that the plants haven't appeared yet. The fields blend seamlessly into the forest floor.

But when he points out where they begin, and how far along the hillside they sweep, I'm staggered. We're standing on top of a fortune's worth of roots—and not a small one, either. Syl has five and a half acres planted. With luck, and a harvest of a hundred pounds an acre, and a steady price of $350, each acre could yield $35,000 in ginseng. And if the plants are allowed to re-seed naturally, this sustainable harvest will continue forever.

Raised virtually wild, ginseng plants require no tending, but Syl keeps busy anyway. Over time, he's been building up lines of deadfall, cutting and dragging the fallen trunks and laying them across the slope to form terraces. The dead leaves are piling up and decaying, building an ever-deepening layer of rich humus.

The beauty of this place is gentle, not dramatic—a deep bowl in a mossy limestone slope, the rounded hills rising behind the trees. Thick leaves are piled soft as rags under my feet. A month from now, the forest will be an explosion of yellow green, but for now, it's only beginning to stir. The thousands of plants he has seeded are awakening, preparing to unfurl their year's new growth, to gather the cool forest sun and fuel their burgeoning roots.

In Syl Yunker's ginseng field, high on the rocky ridge, the only sounds are birdsong and the rush of wind through the bare maples. "I spook a lot of wild turkeys out here," he says. "I've seen 'em so thick it was like water flowing." His boot brushes aside the dead leaves, uncovering the first green shoots of spring.

FIVE

"COME BACK WHEN YOU HAVE A REAL CRIME TO REPORT"

Dr. Terry Jones is the kind of scientist who has actual mud on his boots. When I finally catch up with him in his office at the Robinson Experimental Station, the University of Kentucky's agricultural station in the eastern mountains, he's in jeans and a work shirt, just come in from a project growing blueberries on reclaimed strip-mined land. The berries do well up there, he says. They like the acid soil.

I ask about his research work. "I tend to do fruits," he tells me. "I also look at vegetables. I do tomatoes, peppers, and pumpkins, and we do a little bit with cut hydrangeas for floral markets. Now and then cabbage, broccoli, and cauliflower, but I haven't done those for a couple years. But we look at other crops that people might raise and sell commercially." What he doesn't do is ginseng—not after what happened four years ago.

An older man with curly silver hair and a neat moustache, Jones often looks like he's holding back a wry smile as he tells me the story. Robinson Station's mission, he explains, is to develop practical agricultural techniques that will be of benefit to Appalachian farmers, a group that's had an even worse time than the general run of U.S. farm families. In recent years, the market for tobacco, once an economic mainstay of the region,

---

Previous page: This hunting goods store in rural Virginia was the site of a long-running undercover operation that aimed to put a stop to ginseng poaching.

has been in a tailspin. In searching for a replacement source of income, ginseng seemed a natural choice. So Jones devised a study to determine the best soil chemistry for ginseng, to help growers produce bigger harvests of valuable roots.

"We found the perfect spot," he says. It was out in a mountain forest owned by the university. "We made beds, cleared underbrush, we spent three or four days getting the site all prepared. And what we wanted to look at was whether different pHs had an effect on the germination, growth, and survival of ginseng and goldenseal. So we went ahead and put them on beds, which is going to give a slightly artificial look to the root, because the ground's softer. Plus then the roots are worth less money and we thought maybe somebody wouldn't steal it. I was worried from the day we started, but you have hopes. . . ." A flicker of a smile. "So we prepared the site, and then we adjusted the pH on it."

They planted the same number of ginseng seeds in plots with different soil pH, and waited to see what would happen. Everything went brilliantly. At first all the plants grew well, but within a few years it was obvious which type of soil was the most favorable. All of the biggest plants—three- and four-prongers—were in the same plots, pointing to the optimal pH. This information, if published in a scientific journal, would be extremely useful to growers across the continent. By the fourth year of the study, Jones was getting nervous. In order to get any scientifically valid results out of these observations, he needed to harvest and weigh his plants. He got his digging forks and bags all ready to go. Then the phone rang.

"I got a call one day asking if I'd dug any in my plots. Two of the ladies that work at the forest had walked back in to look at it, and they could see that it had been dug." He drove over to take a look and found his fears had become a reality.

"You just sort of feel sunken," he says. "Your shoulders slump a little and you look and then you think about the time you spent making the beds, and planting, and wishing. . . . The people had come in on an old logging road—you could still see their paths and how they had skidded down over the hill to get it. And what they did was go to the best treatment, the ones with the four-prongers, and dig them. Which destroyed the trial." The smile again.

"Do you have any idea who did it?" I ask.

"No, ma'am. There's a lot of people who walk the forests. . . . They like to be out. When I'm wicked, I say it's all the people on disability and welfare that need the exercise, and they're out and about. . . . No, no idea. Wouldn't have a clue. So for me, there's enough other things to do. It's not that I dislike ginseng, or the woods, it's just . . ." He breaks off, shrugs.

"You sound really philosophical."

"I've had years to become like that. I was upset."

He didn't call the sheriff, although theft of state property is a fairly serious crime, even in Kentucky. "We looked at it like this: How would you find it? How would you identify it? People that live there and worked there didn't have any clue about it, and—I don't know how we'd have gone about it. I just knew I'd been *had.*"

In his work, he still has to deal with people who call for advice about how to start growing ginseng. He tries to give them information without being too encouraging. About the only place that people could hang on to a ginseng crop long enough to sell it, he feels, is on the edge of a town, or in an area urbanized enough that the wildcrafting tradition has been forgotten, and where no one will recognize the plant. "I'll talk to you about ginseng and goldenseal, but I won't tell you to go out and plant it."

He still thinks it may be possible to find land with the right climate and forest conditions for ginseng, and make a success of it. "But away from here. Unless you're surrounded by all your kin and they're mean as snakes." And he is always discreet about what he knows. He is the custodian of the data from Jo Wolf's wild ginseng surveys, along with GPS coordinates for every wild ginseng patch she knows, but he swears this data is thoroughly safeguarded by the chaos in his office. "And if somebody calls and asks about ginseng production, I try not to remember their name."

In the end, Jones thinks there's little to be done about the problem of theft. Better laws and more vigorous prosecution may be of some help, but the fundamental problem is cultural, he tells me. "There's been a tradition of foraging through the woods for plants or materials of value, and that doesn't take into consideration property rights, or the fact that it might be something that somebody else was actually cultivating or working on. I would have paid whoever stole it a hundred times what they got for it, and given them the roots afterwards, if they had just waited."

Ginger Shelby, on the other hand, got good and mad. Mad enough to go out and tend her ginseng field with a rifle across her back. When a bunch of toughs in her tiny Ohio town shot up her valuable hardwood trees, killed her dogs, and stole her ginseng, she picked up her husband's gun. "They want to scare you, to make you bow down, and that makes them continue their threats and intimidation."

When I meet this slender young woman with long, delicate hands at an Appalachian Ginseng Foundation event, her story becomes more believable. "I'm Shawnee," she tells me. "I wasn't raised to be afraid of people."

Shelby's family has its roots in Pike County, Kentucky, and she has vivid memories of a childhood trip to the woods with her grandmother. "She showed me this plant and said, 'Remember this if you're ever in trouble for money. This here's your taxes, this here's your mortgage.'" Her grandmother would take the kids out sanging in the woods, and then, after taking the roots to a dealer, let them pick their own Christmas present from the Sears mail-order catalog.

As an adult, she became interested in the whole spectrum of medicinal herbs and began cultivating goldenseal, bloodroot, and, of course, the most valuable herb of all, ginseng. She got an Ohio ginseng dealer's license and developed a thriving herb business. But then she ran afoul of some of her neighbors, right-wing "militia" members in her small town. After a campaign of harassment and outright theft that grew ever more intimidating, she finally began hauling the gun to her field. But that didn't stop the thefts.

"This is not small potatoes," she tells me. "I lost over $100,000 in my first year. I knew the number of plants and their value. But when I went to the county sheriff, nothing was done. They said, 'Come back when you have a real crime to report.' These poachers are *armed,*" she says. "They'll do anything to get at what they feel is common property."

The gun didn't help, and guard dogs didn't work either. She says she had two Rottweilers shot dead by poachers. Instead, she hit upon another idea: solar-powered, motion-triggered wildlife cameras. She also posted signs all around the property: *Smile, you're on candid camera* and *You are being watched!* "Use simple words," she advises. "Guys like that don't read too well."

With an actual video of a crime being committed, she figures she might finally be able to get more action out of the local sheriff's department. And so far, everything has been quiet—during the winter months she found no tracks in the snow around her fields. Still, she sighs, "It's going to take a long time before I feel safe in my own woods."

As we talk, Shelby is bursting with plans for a hands-on forest museum, ginseng co-ops, a website to market high-value roots directly to China. But when I call a year later to hear how she's faring, all that has changed. She's gone through some harrowing health problems, a brain tumor that ultimately caused a heart attack, and her fields have been completely cleaned out. Her neighbors have gotten her husband arrested on bogus trespassing charges. She's let her ginseng dealer's license lapse, and is now relying on elephant garlic as a cash crop. And she's talking about logging the most valuable trees on her land.

"We're torn," she tells me. "We're getting ready to select-cut, and timber our forest. There's half of us that's angry and says, oh just clear-cut it, and leave them nothing to poach. And I could pay off some of these medical bills. But the other half says, no. That's the part that says—But where will I walk, when I want to talk to God? A forest is more than money. And when they poach, they're not only invading our land, and taking a cash crop, they're stealing our soul. They're taking a part of us."

The sign on the front door of Rock's Dixie Emporium in Elkton, Virginia, reads "CLOSED—Please come back again!" There's another sign in the empty display window of the sagging concrete-block building informing passers-by that the place is up for sale. But it's not too likely that anyone will ever come back again to Rock's, nor is anyone going to buy the store anytime soon.

High up on the faded yellow front wall, tall black letters tell the world what Rock—yes, there once was a Rock here—used to do for a living. "Big game check station—Civil War artifacts—Specialist in ginseng." The assortment of enterprises only seems odd if you don't know that Elkton, with its six thousand people and three traffic lights, stands at the edge of the Shenandoah Valley. It's a stone's throw from Civil War battlegrounds and a handful of miles from the entrance to Shenandoah National Park, some of the most beautiful and threatened wilderness in the entire eastern United States.

The two-story building stands vacant and dusty now, but during the years that Rock was there, it was a very different scene. A steady stream of customers came and went, some buying and others selling, trading in wild ginseng by the pound, coming in search of black bear paws, black bear gall bladders (believed to be another miraculous healer in traditional Chinese medicine), and even of whole, dead black bears. Rock welcomed them and cajoled them, dickered prices and gave advice.

I try to picture how stunned his customers must have been to learn after two years that Rock, big, folksy ol' Rock, was not some nice old fellow who retired to the mountains to lose money running a hunting supply store, but an undercover federal agent. No one guessed, until the arrests began.

To hear the story of Operation VIPER, the most successful attack ever on ginseng poaching, I've come here to the Blue Ridge of northern Virginia to meet with one of the people behind it—Skip Wissinger, a special agent for the National Park Service (NPS). On a sunny morning at the very end of autumn he picks me up at the park's Swift Run Gap entrance, two thirds of the way down the Skyline Drive's 105-mile north–south run, and I climb into his dark blue Chevy Tahoe with its white government plates. He's a trim, deliberate man in his fifties, dressed in black jeans and boots, with the tip of a leather holster peeking out from under his green sweatshirt, just in case anyone thinks he doesn't mean business.

It occurs to me as we head off that I should have come a month earlier—the fall colors draw traffic-jam mobs of visitors from Washington, D.C., just a hundred miles away—but the brown lace of branches has its own beauty, and the view is even more stunning through the bare trees. I look down over the Shenandoah Valley, and beyond to row on row of mountains fading into blue. It's old country here, America's very first west, settled in the 1710s. The barns in the valley are painted white, and the big, venerable houses have a brick chimney at each end, thin tails of smoke unfurling from some of them. Up here on the ridge, winter is on its way. Already plumes of ice trail down the rock at the side of the road.

So how does a park ranger in such an idyllic place end up a criminal investigator? "Traditionally," he tells me, "the NPS has had a park ranger position that deals with all kinds of protection—Emergency Medical Services, wildland fires, search and rescue, law enforcement." Over the years, a need arose for someone to focus on long-term investigations, and the parks created positions for special agents, who actually hold the same federal job classification as agents for the FBI or the Bureau of Alcohol, Tobacco, and Firearms.

Naively, I wonder what kind of crimes there are to investigate in these unspoiled chunks of nature.

"It depends on what's happening," he tells me. "For a special agent working in Arizona, like Organ Pipe, a major issue there is drug smuggling. If you look at Grand Canyon, you'd find a lot more people-oriented crimes being investigated, assaults, rapes, and that sort of thing. There's three of us at this park, and one person's time for the last two years has been occupied dealing with a double homicide here a number of years ago, of two female hikers. Carrying that through to a prosecution, which can be very lengthy. So here I am dealing with ginseng and bear parts, and the person next to me is dealing with a homicide. The only thing we don't investigate very often is bank robbery, and that's because we don't have any banks here. But there is a bank out at Grand Canyon. So they probably have to deal with that!"

We pull off the road at a glorious overlook, and Wissinger kills the motor. "Just how bad is the ginseng poaching problem here?" I ask him.

In Shenandoah National Park, he tells me, around 80 percent of the ginseng that should be there has vanished—stolen. In recent years, the NPS has been working with predictive mapping to locate the spots where ginseng should be growing. This is similar to the work done by Dan Bond of the Appalachian Ginseng Foundation, pinpointing prime ginseng-growing land for farmers. In the National Parks, collection of any plant or animal species is strictly forbidden by law, so given an accurate computer model, ginseng should be found in all the habitats where it is predicted. But it's gone.

Everyday experience as a ranger backs that up—Wissinger remembers how common the plant was decades ago when he started at the park. "And also simply looking at the amount when someone is apprehended," he says. "If they had twenty roots in their possession, and they've been out all day, they're usually the first to tell you—'God, I worked my bottom off to get this, and it's not anywhere like what I used to get.'" He starts up the Tahoe, pulls back onto the deserted road again. "We have had poachers that have traditionally taken ginseng from this park, they live here at our back door, and they have made statements that they no longer come here because it's not good digging anymore. They travel to West Virginia because the poaching is better there."

The first investigation into poaching in Shenandoah took place in the mid-1990s, when Operation SOUP (Special Operation to Uncover Poaching) revealed the existence of a booming international trade in black bear parts from Virginia that were being shipped off to Asia for use in traditional Chinese medicine. The investigation concentrated on the poachers themselves, and an extremely brave undercover agent actually managed to infiltrate hunting groups and go out hunting with them. Along the way, he learned that many of the same people were also trafficking in stolen ginseng.

"The other thing that was learned," Wissinger tells me, "was that there are different levels of structure within the black market. Think of it as a ladder. The lower rungs on the ladder are suppliers, and then at the other end is the consumer. And then there were multiple mid-rungs of people who were buying and selling this product, not as consumers, but to make a profit. So we were aware that you could attack the problem from several different levels. And from a strategic standpoint, we felt that focusing on the consumer end and the midlevel profit-making part of the black market chain we might have a longer-term impact. Because the ultimate goal for us was to have our roots in the ground here in the park, untouched." By destroying the marketing apparatus, they hoped to end the trade.

Together, the National Park Service and the Virginia Department of Game and Inland Fisheries planned an investigation to penetrate the network of buyers and middlemen responsible for plundering the park. It was named Operation VIPER—Virginia Interagency Effort to Protect Environmental Resources. "Who comes up with these names?" I ask Wissinger. He laughs. "Different people. Always an acronym!" Ultimately, the investigation also involved the FBI, the National Fish and Wildlife Forensic Laboratory, the U.S. Attorney's Office, and Virginia state courts.

In the state of Virginia, it is illegal for anyone to buy wild ginseng without a dealer's license, so everyone who purchased roots from Rock, whether for personal use or resale, was committing a crime. It's also illegal to dig ginseng in national parks, and to dig or sell outside the authorized season. And needless to say, it's illegal to traffic in the body parts of black bears. By setting up a shop and carefully recording every illegal transaction that went on there, investigators hoped to get an inside view of the scope and operation of the illicit market in plants and wildlife.

By the time Rock's Dixie Emporium went out of business, the operation had documented 99 federal felonies, 105 federal misdemeanors, 193 state felonies, and 294 state misdemeanors, committed by over 100 individuals from seven states, the District of Columbia, and one foreign country—691 criminal acts in all, witnessed by Rock, whoever he is.

The next place we stop at is a wooded area with a huge fallen tree trunk sprawled among young saplings. It's just about the prettiest crime scene you can imagine. "One of the cases that we investigated with Operation VIPER," Wissinger tells me, "is a person who wanted to bring a friend of his to the park and buy some ginseng. And they were very concerned about authenticity. So the deal was set up, and the agent met with these people and they came to this very spot where we're parked, and just beyond that log in the woods there were several ginseng plants. This one fellow flew here from Korea, and just seeing a plant growing added authenticity to what he was buying. The plants that are here are marked—he actually pulled one while he was here! And he took that plant, top and all, back home with him, along with several pounds that he bought. Literally the next day he was on a plane on the way back to South Korea." Consumers were starting to ask for Blue Ridge ginseng, even Shenandoah ginseng, as though it were a brand name.

Every bit of the ginseng that was sold, he reassures me as we continue down the road, was brought in by poachers—park personnel didn't go out digging up the park's resources. The bear parts were a little trickier, since no one wanted any bears getting killed. "Most of our whole bears that we sold came from roadkill on the highway in one state or another, and the states helped us, provided bears for us. I remember driving to Maryland and Pennsylvania. And of course, those bears had to be kept in a freezer," he laughs. "We had to maintain our inventory—you can't just go off to Wal-Mart and say, 'We need two bears instead of one!'" Just then, a buck with a handsome rack of antlers strolls into the road just ahead, takes a long and thoughtful look at us, and wanders off into the trees. "He's a good-looking guy," says Wissinger.

Now we're in the center of the park, in Big Meadows, a ranger station and maintenance area that's the district office for the central section of this park, and Wissinger pulls up in front of a low concrete building. "This is the investigative office that we work out of. And this is where Rock is supposed to meet us." That's news to me—very welcome news.

It looks like a very well-worn motel, and in fact it once served as housing for seasonal employees. Behind a door that's still marked "Apartment Six" I enter a string of offices jammed wall to wall and floor to ceiling with desks and computers and shelves and file drawers and big metal cabinets.

"Let me show you something back here," he says. He's standing in front of an opened cabinet filled with fat, white binders, floor to ceiling. "These are the actual case files. Let's just pick one, to show the work that went into one of these."

He pulls out a four-inch-thick binder and opens it to reveal a mass of documents, neatly sectioned by dividers. "This is what would be handed to a prosecutor. An overall synopsis of what took place, a transcript of every telephone conversation between the agent and the person calling, word for word, a report of an illegal sale, that took place on this date, here are the names of the people involved. Each of these are a different date. Word for word. We kept a transcriber busy, well-employed. Here's a report of Rock's on the transaction."

Next comes the evidence section, with a written record of every piece of evidence, what it was, where it is, the numbers to track it with. "Here's a photo of the person's vehicle at the store, at the time, this is them leaving. Here's evidence that was sold—gall bladders, and some roots—and then the chain of custody, showing the flow from the forensic lab. Everybody signs off, and it doesn't change hands without it. Here's a laboratory report, an examination that shows that this particular item is a black bear gall bladder. Then you get into another section listing all the people involved. Then a listing of all the charges on each date—again, this is prepared particularly for the prosecutor, a list of all the government witnesses that might need to be subpoenaed."

All this is for just one case?

"Yes, ma'am! And this is an indictment form—this case has actually gone forward in state court, in Rockingham County, and this is the formal charging document that charged this person: 'On or about the fifth of January 2001, in the county of Rockingham, Myong-Chul Kim did unlawfully and feloniously purchase a wild animal in violation of—'"

Kim, Kang, Chang, Lee, Kim, Kim . . . I read along the rows of binders, all labeled with the names of defendants. Every single one of them is Asian—more than a hundred names.

The phone rings. "Investigation, Skip speaking." A loud voice garbles through the receiver. "Hey, I was afraid I had missed you! Yeah, put the pedal to the metal and come on!" He beams, then hangs up. "That was Mr. Rock. He's the county game warden east of here now, in uniform, so he's not that far away. And he has some things he had to bring back up here."

"How did he end up being the one to run the store?" I ask. "Did he volunteer?"

Wissinger grins. "Have you ever heard the story that they lined everybody up and they asked the volunteers to step forward? Rock's mistake was that he stood still, and everybody in line with him stepped back one step. He's a type-A personality. And he's not short on opinions. And, on a personal level, we just love him. Talk about dedication! What he has done is so above and beyond the call of duty . . ." He shakes his head. Then he rummages through a file and pulls out a computer-printed photo in a plastic sleeve. "That's the store, the interior picture. That's basically what it looked like. And that's Rock."

Behind a tidy display counter leans a burly, gray-haired man in a plaid shirt, unsmiling, his hands a big heap on the counter's edge. The store is decent-sized for a one-man business, well-organized and brightly lit by long fluorescent tubes. Around him are deer heads mounted on the walls, a rack of fishing lures, shelves of ammo and videos about hunting. The clock on the wall behind him reads 10:45, and he's been drinking something from a tall plastic cup.

Another photo, an enormous paper sack with its top rolled down to show it is full of roots. "That's ginseng. A significant amount of it came from the park, and it was sold commercially in D.C. What the person bought was ten pounds out of that bag. And—" He lays on the table a grainy surveillance photo showing a back-view of a dark-haired man in a ski jacket stowing two items in the open trunk of a car. "The sack has ginseng and in the cooler is a bear gall." He won't tell me how the photo was taken, because this case hasn't come to court yet.

It's nearly a year since the investigation was wound up, and the more than one hundred cases are all in different phases. Of them, about half have gone to federal court, and the rest to the state of Virginia, depending on the nature of the offense. Out of all of those cases, he says, 90 percent of them never actually go to trial—they are resolved through plea-bargaining. Because state courts move a lot faster than the federal system, most violators under state law have already been charged and convicted, of a felony. To date, all but one of the people charged have been convicted.

And what happened to them?

"They've paid a restitution amount that varies according to what they bought—generally speaking $1,500 per bear gall bladder, or $1,000 for a pound of ginseng. And if there's three defendants, it's divided three ways. So we've had people that have paid restitution of as low as $1,500. And actually when you split it out, and there's two or three people together, we've had people that have paid as little as $250, taken a felony conviction, and had a suspended jail sentence of one or two years."

"That's all?"

He shrugs.

"What had you been hoping for?"

"More than that. . . ." He pauses, starts to say something, changes his mind. "There are some areas that are a little difficult for me to talk about, because we're in the middle of an investigation. I'm only a couple of years away from retiring, and I'm not joking when I say I want to make sure this thing gets finished. It just moves so slowly. . . ."

The outer door bangs open. "Hey, bud!" a voice growls from the next room.

"How are ya?" Skip calls.

"I got here!"

"Rock, you're a good man. You're so famous that this lady drove all the way—"

In walks a stocky figure in a bright green uniform and a bright green cap with a big green cloth sack slung over his shoulder. He looks like Santa Claus gone color-blind. "I got everything you need here. The ginseng too."

Rock, I later learn from his business card, is actually Virginia State Game Warden Rocco Cianciotti. The sack contains a collection of items confiscated as evidence during Operation VIPER and now being used to train law enforcement personnel who wouldn't know a 'sang root from a bear gall bladder. Rock is bringing it back from a conference he attended.

A minute later, as they're sorting through the sack's contents, Wissinger calls to me from the back office. "You oughta see this, too." Spilled out on a desk are a jumble of objects tagged with fluorescent green labels that read "EVIDENCE."

"This is a bear gall. Have you ever seen a dried bear gall?" No, not really. It looks like a deflated black balloon, six inches long and dry and hard as leather, with an odd, spicy scent. "That one is a little bigger than medium size, and it's just been hung and air dried. It doesn't exactly blow my hair back."

"And here's some ginseng that's been marked," says Rock. He proudly holds up a much-handled plastic bag filled with dried roots. There's nothing unusual about them, except for faint rusty discolorations in one or two places, easily overlooked. "Just wait!" The two men rummage through drawers and cupboards, finally rustling up a black light the size of a flashlight, and someone flips off the overhead light. "See that?" Under black light, the invisible markings scream orange like the sunset in a psychedelic poster.

The method was developed by Jim Corbin, a biologist for the North Carolina Department of Agriculture, to identify and protect wild ginseng growing on protected land. The dye is made up of calcium and magnesium compounds, along with tiny silicon chips coded with the exact location of the plants. Though poachers may scrub it off the root's exterior, the dye is absorbed deep into the plant and is easily spotted under black light. Further refining the method, Corbin spent two years training his dog, a Doberman, to sniff out the dye, with great success. In their first bust, the dog detected a single marked root inside a 450-pound barrel of ginseng. That dog has assisted in 140 seizures of marked roots, and Corbin has since trained a second dog. So far, over 22,000 plants in the Great Smoky Mountains National Park have been tagged, and poaching has fallen off drastically. Clearly, Corbin's method shows great promise for protecting off-limits plants in National Parks. Chinese buyers, however, may be less partial to legally grown roots embedded with microscopic tracing-chips.

Besides the day-glo roots, the two men show off a whole plastic bag stuffed full of gall bladders—each one cost a bear its life—and a silver-and-turquoise bracelet studded with bear claws. Most horrifying of all is a pair of buckskin necklaces, with the preserved foot of a golden eagle, talons curled and grasping at nothing, hanging from each.

All this, Rock tells me, was evidence in a single case, which has since been adjudicated. "The man who's selling these is from Pakistan. And all that jewelry you see, he has it made by a Native American. He does it for profit."

What happened to him? Has he been sentenced?

"He did a plea bargain in federal court. Got his hand slapped," Rock says curtly. "Anyway, that's some of the stuff we netted out of the ginseng investigation. We have a few thousand pieces of evidence."

"I think you'd look pretty good with that on!" Wissinger teases.

Rock holds an eagle necklace against the chest of his uniform. "That right there, retail on that, the violator said it was somewhere close to $2,000. Now these, I would *not* want them to disappear!"

"They're not going to disappear," Wissinger assures him.

"I know how law enforcement is. They *improvise.*"

Skip laughs hugely. "I think if someone *improvises* with that, they're going to stand out!"

I try to visualize one of these guys at a party with a purloined eagle claw dangling from his neck—I don't think Rock needs to worry about this evidence getting "lost."

"You must have had an interesting time in that store," I say, as they pack the evidence away in a cabinet.

"A forgettable time," says Rock, flatly. "It wasn't my cup of tea." But the story comes spilling out anyway.

Rock, Rocco Cianciotti, grew up in Missouri and joined the Marine Corps the day he turned eighteen, serving for four years. At the end of his enlistment he was stationed in South Carolina, and recruiters from various big-city police forces came around to talk to the short-timers. The Washington, D.C., Metropolitan Police Force had the best deal going—a good starting salary, and retirement after twenty years—so that's where he signed on. "I left the gates of Parris Island and drove straight to Washington, D.C., and went through the revolving doors and upstairs to the third floor, me and four other military personnel, and they gave us a gun, and a cap plate, a baton, six bullets, a badge, and an ID folder, swore us in, and said, 'Go home and practice dry firing a pistol until the academy class picks up. You are now a metropolitan police officer.'" It was 1969.

After five years as a patrol officer, he moved into criminal investigation, and ultimately spent twenty years as a homicide detective. "I adopted the view that, I work for the Metropolitan Police Department, but I serve the dead. The reason being, no matter what awful crime is committed against a human being—like I could never work the sex squad, because I could never handle what happens to the children—you're still alive. You can communicate with somebody what happened to you. But some person lying there that died at the hands of another can't do that. So when I walk on that scene, I'm gonna speak for him.

"And that means that when I look at him or her, from that moment on, I think *prosecution.* What could I do wrong in this step where they could rip me apart in court, and not make your perpetrator serve the time? So I carry that over in everything I do."

But working homicide told on him, and he'd always had a passion for the outdoors, and at age fifty-four, ready to retire from the police force, he spotted an ad for game wardens for the Virginia Department of Game and Inland Fisheries. Thirteen positions were available, and 690 people applied. He got the job and headed back to the police academy for twenty-two weeks. And when he graduated, his new employer was looking for someone to run an undercover operation. They needed an agent to track down wildlife traffickers around Shenandoah National Park.

The store in Elkton had already been opened, but it wasn't getting any results, though neither Rock nor Skip wants to discuss the reasons for that. Rock decided to start over, from scratch. "I talked to people in the neighborhood and found out who owned the premises, instead of going straight to it, and I let the word go round that I was looking at several other locations. My covert background had taught me that, no matter where you're working, whether it's in the inner city environment or out here in the backwoods country, there are certain things you always do. And one thing is, you don't change who you are, and you don't prefabricate too much. So I kept my own name, in case anybody I knew ever bumped into me. And I went and applied for all the licenses, the ginseng licenses, the business licenses, we filed our quarterly taxes. It looked like a legitimate transition." His cover story was that he had retired from thirty years as a civil servant in Washington, and just wanted to run a nice, quiet little sporting goods business, there at the foot of the mountains. And it worked. People began coming in, checking him out.

"I knew that the assignment was going to be somewhat labor-intense when I took it—but I had no clue what it was going to turn into. No clue," he says ruefully. "When I went into this, I knew nothing about ginseng. So they gave me a three-day crash course in a motel room. Now mind you, I've never seen ginseng in my life! And we have this buyer on our side, a big heavy dude with a beard—" He imitates an ornery, impatient man with a heavy mountain accent: "'Now this here's wild, and you know how you can tell, you see how tight the rings are and they curve together? Now this ginseng here is transplanted, you see where the rings are tight and they start to expand? And this ginseng here is cultivated. You got that, boy?' That was *it!* And I think, Shit, and I've gotta *sell* this stuff?"

The agencies expected him to work regular business hours, but the reality wasn't quite so simple. Anytime anyone wanted to do a deal, he had to be there in the store. "Social life—I had no social life," he complains. "It was such an old, decayed, *delapilated* building, and I started worrying about my health, because the mold, the stench, and—oh Jesus, it was horrible, and you have to live there! I had to secure it, lock it up, and I couldn't bring people in because I had all the electronics upstairs. In our store, we had a clock in the back where everybody could see it, just like you do in any store, and hidden in the number 4 was a small camera. I had all the equipment that would make it work upstairs." This was how he lived, for two solid years.

Along the way, he learned a vast amount about ginseng. At first, he had to find out market prices by calling around to other dealers, pretending he had roots to sell. Friendly diggers started sharing their knowledge and experience with him, showing him how to tell when ginseng had been frozen, or salted with various other roots. When he offered to buy newly dug "wet" roots, a man he'd befriended told him that he was getting ripped off, paying good money for the dirt that was still clinging to the roots. He installed a sink in the store and made diggers wash off fresh roots before he would buy them.

And all the while, the video camera was watching, set off whenever a transaction was about to take place. The tape would roll, the deal would be done, and afterward Rock would spend hours logging every detail of what had occurred, his eye always on that faraway moment, years in the future, when the case would finally go to court. Now there are shelves and shelves in this office filled with the tapes.

"Would you like to see one?" says Skip.

"Can I?" I ask, surprised.

"Sure, long as the case has been adjudicated. . . ." Skip goes into the back room and rummages around, and the two have a long, shouted discussion about case numbers and where they might be. Finally Skip turns on the television that sits on a shelf, pops in a tape.

It's not at all what I've expected to see. Raised on "Dragnet" and "Mannix," I'm looking forward to shocking revelations, tortured looks, passionate words. Instead, there are three timid Asian men, seen over Rock's shoulder, fiddling silently with something on the counter, and Rock droning on about the best method for storing fresh ginseng (keep it in lots of paper towels, and squirt it with a spray bottle of lukewarm water—never cold). It's all in glorious black-and-white, not an ounce of drama, and acted at a snail's pace. Suddenly I get a glimpse of the glacial steely patience it must have taken to run this operation. Two endless years of this. Skip hits EJECT, knowing I've seen enough.

I finally ask Rock a question that's been nagging at me. "If all these people were Asian, and half of them didn't speak English, how could you warn them that this was illegal?" This is without a doubt the most controversial aspect of Operation VIPER. Every single one of the buyers charged in the different cases was Asian, and the vast majority were Koreans or Korean-Americans. Immigrant rights advocates whipped up a storm of criticism with their charges that the buyers were unfairly targeted because of their cultural practices, that they had no way of knowing the illegality of what they were doing, that they were led straight into an unjust trap.

"With my law background, I know what entrapment is," he says, shoving his green uniform cap back at an obstinate angle. "Entrapment is not a law or a statute—most people don't know that. So you have to be kind of innovative. How would anybody ever know that these people actually were warned that the deal was illegal?"

His undercover years obviously made him a talented actor. For my benefit, he replays all three roles in a typical deal: himself, the buyer, and the buyer's English-speaking friend. "I would say, 'Wait a minute, before you walk away from my store with anything, I need to know if you understand what the hell you're doing could put us both in jail. And I ain't showing you nothing until I know you understand that!' And then one would look at the other and say, 'What's he saying?' And they'd go ohhhhhhh!

"And I'd say, 'You understand?' And I'd take their hands and make like handcuffs—'You understand?' 'Ohhhh! Jail, oh yeah, jail.' Things of that nature. And I'd look at the English-speaking one, and I'd say, 'Hey, are you sure this clown knows? Because what's gonna hurt me is if he goes out of here running his mouth, and then you're gonna put the man on me!'"

It's a strange, complex character that he's portraying, a jovial good ol' boy with a hard urban edge about a millimeter below the surface. Rock was born and raised country, from coal-mining people, he tells me, but left that far behind him. "I didn't try to be one of the locals," he says. "Like if one of them dumb-ass country boys come in and start talkin' shit, I'd look at them and say—" this comes out in a well-worn litany of menace—"'Look I don't joke smoke dope skip rope, I quit school cause they had recess, I'm a *bad* motherfucker, I just don't *play,* you understand me?' And they'd look at me like—'*What the—?*' So I didn't try to be one of them."

Behind it all, he says, is greed, pure and simple. Even the most suspicious local who came in to sound him out would, sooner or later, let greed overcome canny judgment and end up doing illegal deals with him. He understands that mentality, he tells me, because he grew up in it. "Ginseng is a quick fix for a lot of the poor people that never had a chance to elevate themselves. And a lot are very uneducated, and they know this works. They work steel, they cut trees, or work in a coal mine. And you never get out of that circle. A lot of old-time diggers, and even the young ones, they're of the same mindset, and they actually don't see it as hurting anything. They look out and they see this vast land that we have, and they say, 'What does it hurt?'" He imitates their puzzled looks.

"They don't understand, because they're uneducated. But if you let it go unchecked, and let it run rampant, you could deplete our natural resources. To the point where your grandchildren will never see this."

So, I ask as he stands in the hall, shouldering his backpack and ready to leave, how do you feel about how this has all turned out?

"Honestly?" He looks at me sideways.

Are you satisfied with the convictions you've gotten?

"I can't say that I am. With the amount of time that went into this, and the attitude of some of the prosecution that I see—no, I'm not happy with it. Not at all. And we haven't even scratched the surface of what I've learned, from those diggers and those dealers. . . ."

But he's not a man to waste time brooding, and he has to get back on the road. He takes his leave of Skip, then turns back to ask me what I'm going to write about next. First I have to finish writing this book about ginseng, I tell him.

"Well, if you ever wanna write another book," he says, "you should write about me from how I come up to how I got here. I mean, I worked undercover on the Mayday demonstrations,

I've met presidents, I've toured with the director of the Bureau who used to be a homicide detective—I've led a very colorful life. I swear. . . . If you ever want to write a book like that, you look me up! And sometimes I think back on all the stuff that we did, and I laugh out loud. I mean, it would be a comedy of tragedies." And with that, he's out the door.

Later, as Skip drives me back to the park entrance, I ask why this massive operation has had so little real-world effect.

Part of the problem, he tells me, is just getting prosecutors to take these cases seriously. Depending upon the exact nature of the offense, some of the cases have been prosecuted under state law, and others under federal law. The state court system moves much faster, and most of the cases filed there have already been settled. Trafficking in ginseng and bear parts from a national park is a more serious offence under federal law—especially if it involves interstate commerce. But the federal court system is overburdened with much more serious cases, and it can be difficult to even get prosecutors' attention.

"When I walk in with a case file to a prosecuting attorney," Wissinger tells me, "and I say we have this case and it's ready for prosecution, I'm competing for prosecutor time with the FBI dealing with a bank robbery, an ATF agent that's dealing with serious gun trafficking, the DEA dealing with drugs, and who knows who else dealing with embezzlement. And they turn around and look at you—'What's ginseng? It's a dirty *root?*' There's some joking, and I'm sure there's some snickering in the other room after our folks walk out—'a piece of gall bladder? from a *bear?*' And the question—*Is that a real crime?* That comes up. And I think, if you're at all environmentally connected, and I'm speaking from the heart here—yes, that's a real crime."

And ginseng is even harder to prosecute than bear parts. "A bear is warm and cuddly," he says. "Compare that to a dirty root. That sort of aesthetic appeal going into this doesn't make it any easier. If nothing else, we have raised the level of awareness that ginseng is being trafficked in the black market and to a more extensive degree than we otherwise would have guessed."

Ultimately, what will be the outcome of Operation VIPER—of Rock's lonely two-year stakeout, of the countless thousands of hours of paperwork, the prosecutors carving a hole in their jam-packed schedules?

His smile is pressed very thin. "The prevailing wisdom is that something like this will make a dent, and then in a year or two the poaching will start creeping back."

But it's a beautiful afternoon, and we're traveling the Skyline Drive, a prime scenic destination, and it's hard to stay drowned in gloom. Instead we start talking about how Wissinger came to this dream job—a National Park Ranger in one of the world's most beautiful places. Though Shenandoah's campgrounds shut down in the fall, the Drive is open all winter—unless it snows, hard. And that's one of the truly magical aspects of his job. To get to work after a snowstorm, he has to drive through miles of unbroken, pure, white, trackless snow, past trees frosted deep and rock cascading with frozen miniature waterfalls.

He tells me about his college summer jobs in the Pennsylvania state parks, and how he landed a seasonal position at Shenandoah one year, curious to see how the National Park Service compared. He fell in love with the place. That was three decades ago, and he still hasn't come to take it for granted.

"Every so often when I'm going somewhere, I'll stop at an overlook or two and just soak in this area. And when I get to the office, I tell the folks up there, I would have been here sooner, I took five or ten minutes to stop and look around, just to remind myself why I come to work every day."

He drops me off at Swift Run Gap, then turns around and drives, slowly, back toward the crowded cinder block office. Another case is coming up in federal court, two weeks from now.

What would it take to put an effective end to ginseng poaching? Clearly, a first step would be enforcement of the laws that already exist. Gathering of plants in national park land is always against the law, and with the invisible marking system developed by Jim Corbin, it becomes much simpler to prove that roots were illegally harvested. Obviously, it's impossible to mark every root on every acre of every national park where they grow, but the practice should also have a deterrent effect on would-be poachers.

But, as the outcome of Operation VIPER has shown, even bringing national park cases to court may not result in a sentence that's heavy enough to make a difference. Prosecutors are unlikely to see the importance of a "dirty old root," especially against the daily backdrop of murder and mayhem that passes through federal courtrooms. Education and greater awareness of the problem may help—to some extent.

Away from the national parks, the situation is so murky as to seem nearly hopeless. As things stand today, there is absolutely no way to prove that any ginseng root, whether wild, virtually wild, or cultivated, was obtained legally. Theoretically, diggers need permission to harvest anywhere other than on their own land, but there is no mechanism to certify this, and no way for honest dealers to find out where diggers got the roots other than taking their word for it. Buyers of ginseng have to obtain licenses for every state in which they do business, but diggers don't need a permit, unless they want to harvest in national forests.

And even when a poacher is caught in the act, on private land, it can be difficult to do anything about it. "There's no way to trace ginseng," says Bud Baumgardner, a deputy sheriff in Hardin County, Kentucky. Because the crime is nearly impossible to prosecute, law enforcement does not want to waste energy getting involved.

Gary and Beth Anderson, two growers in Kentucky, caught poachers on their land more than a dozen times, and the thieves always insisted that the roots in their pockets had come from a neighbor's place. It was only when the Andersons managed to get an officer out to their place in time to arrest the intruders still in their field that they were able to press charges. The case finally went to a jury trial two years later, and the poacher was convicted and fined $1,500—with no compensation for the valuable roots that were dug up.

And what about virtually wild ginseng, which looks identical to wild but is deliberately planted and tended? The ever-resourceful folks at the Appalachian Ginseng Foundation have come up with a proposal based, oddly enough, on the marketing of tobacco—once a significant crop throughout the heart of ginseng's natural range. Syl Yunker and others believe that a marketing permit system like the one used for tobacco could effectively end the poaching problem. Currently, farmers growing tobacco receive a "marketing card," a permit specifying the number of pounds they may sell.

A similar program could be used for ginseng, not to control the size of the harvest (the purpose of the tobacco permits) but to certify ownership of the roots. After inspection of a field, the grower would receive a card entitling him or her to market that crop. Licensed dealers could legally buy only from growers with a card, putting the local poacher out of business. The AGF has had interest expressed by various state legislators, and one Kentucky congressman was intrigued by the idea.

In the real world, there have been some encouraging steps taken against poaching. Rural Action, an advocacy organization in Appalachian Ohio, has held interagency meetings with law enforcement officers to promote the understanding that this is not a trivial issue—many thousands of dollars are at stake. In general, says Scott Persons, legal action is best taken through wildlife protection agencies rather than sheriff's departments. "Wildlife people see it more like poaching a deer, or dynamiting fish," he says. "They see theft of ginseng as a legitimate part of what they're supposed to do. Plants have gotten into their consciousness, where it used to be fish and game." And Jim Corbin's marking techniques make it far easier to bring a case to court, since there is clear evidence of where the roots came from.

But for now, the insoluble problem of poaching has put a bitter end to many ginseng dreams. In the words of Dr. Terry Jones, "It's made some people rich, it may save the forest, it may save human life. It's a crop I'd like to deal more with, it's just—I can't stand the heartache of it."

SIX

FROM THE HOLLOWS TO HONG KONG

In his store's old-fashioned plate glass front window, easily ten feet high, stands a set of scales so venerable that no one remembers exactly how old it is. "It looked just like that when I was a boy," Steve Goodman tells me, and he's in his sixties now. It's a balance with a low curved pan and a set of flat weights, a pound, two pounds, all made of the same gray, utilitarian metal. In the last century or so, it has measured out untold millions of dollars worth of wild ginseng roots.

Behind the scale, red letters on the tall window spell out "S. Goodman and Sons GINSENG—GOLDENSEAL." I'm visiting the oldest dealer in ginseng and medicinal plants in the Commonwealth of Kentucky. Most days in the fall, two or three times, the doorbell will ring in the time-warp front office announcing a ginseng digger who's come to sell his harvest, and Steve, or his wife, Linda, or their son, Seth, will unlock the high, arched glass doors. It's a simple transaction. Goodman looks over the roots, weighs them on that ancient scale, and unless there's something obviously wrong with the ginseng, he pays out a standard price. There's no haggling, and no grading involved—he tells me ginseng is all the same until it reaches Hong Kong. He also punctures the myth that the Chinese are looking for roots

Previous page: The Goodman family of Louisville, Kentucky, dealers in ginseng and medicinal plants for five generations, in front of their warehouse.

that look like a human body or anything else—in thirty years, he says, he's never seen them show any interest at all in the shape, no matter how peculiar it may be.

The storefront is sparsely furnished. Apart from that wonderful scale, it holds a couple of battered wooden tables (one heaped high with empty cardboard shipping boxes to re-use), an old office desk, and ancient wooden chairs. Unlabeled sacks and a couple of cardboard drums conceal their exotic contents. The air smells faintly earthy—though that may be just my imagination—and the only sound is the generic rumble of construction noise that filters through from the street.

Goodman leads me into the back office, and I can't help thinking that a lot of executives would kill to have a suite furnished with century-old pieces like these. There's a long, elegant table for meetings, a pair of wooden chairs with their backs curving in graceful arcs, a tier of wooden pigeonholes stuffed with old cigar boxes and rubber stamps and gizmos whose proper use was forgotten decades ago. For the Goodmans, these aren't precious antiques. It's just the office furniture they've always had. The room feels as cozy and mismatched as Grandma's sitting room.

Steve graciously fetches me a cold drink, then rummages through various cupboards and drawers to bring out photos from the firm's glory days in the early twentieth century. S. Goodman and Sons was founded in 1875 in Glasgow, Kentucky, in the south-central part of the state, and moved here to Louisville in the 1920s. For over a century, the Goodman family traded in furs, wool, cowhides, and medicinal plants—all products from the countryside that they bought into their city warehouse. For sixty-seven years, they operated out of a four-story pre–Civil War building on Market Street, in central Louisville's warehouse district. A photo from the 1920s shows the staunch brick facade, with WOOL HIDES FURS ROOTS painted in bold, vertical letters between the second-story windows. Once, Goodman tells me, there were four or five such companies on this street. Today his is the only one remaining, now in a smaller building, nearly as old, just a block away.

The doorbell rings, and the sound of footsteps and women's voices echoes around the high, wooden rooms. A minute later Linda looks in. "Steve, we've got some customers." He excuses himself, and I tag along. In the front shop two slender, middle-aged Chinese ladies in blue jeans stand beaming at a photo album of the Goodmans' grandbaby, Shayna—the sixth generation. "She look like a boy," one of the women says. "So next baby will be a boy. Some Chinese people believe that." Everyone laughs.

Steve brings out a metal container the size of a popcorn tin and pops its lid off. The women pounce on the dried ginseng roots inside, tossing some decisively into the weighing pan of the venerable scale, rejecting others. "Kinda small," says the more talkative of the two, frowning. "That's all you have? This one is not as good as the one my sister got." They continue the discussion among themselves in Chinese, but it doesn't seem to dampen their enthusiasm.

Meanwhile, Linda has seated herself at her wooden desk to watch the proceedings. She has a gentle face, fringed with soft, brown hair. In front of her stands an old-fashioned banker's lamp that looks like it dates from the first days of electricity—and a bright blue Apple iMac. She tells me it took years before they finally gave in and bought a computer. S. Goodman's first concession to the modern world was a copy machine, which, they had to admit, did make life a lot easier. Some years later, she finally talked Steve into getting a fax machine—but only because a major customer in Germany balked at waiting for paperwork to travel back and forth by mail. At last, they were forced to give in and buy a computer when they discovered, to their dismay, that correction slips for typewriters were no longer being made. "This family isn't too big on change," Linda laughs.

The Chinese women finally settle on three pounds of cultivated root, and Seth packs it carefully into a paper grocery sack, then puts that in a plastic bag. "That's all you have?" one asks. "I can get you more," says Steve, "but give me ten days." She pulls out a black leather wallet fat with bills, and they head for the back office to do the paperwork.

"How many times you write down my phone number?" she complains good-naturedly.

"Every time you come in here," says Steve.

"I come in for twenty *years!*" she protests.

S. Goodman and Sons stands in a row of staunch, 150-year-old brick warehouses, just a few blocks from the Ohio River. That river is the whole reason Louisville is there. Centuries ago, the rough water of the Falls of the Ohio were an obstacle to river travel, and a settlement was built alongside them. Louisville became a major shipping point on the Ohio, the country's inland artery between Pittsburgh and New Orleans, and goods of every sort were shipped south. In the old days, when overland travel in the mountains was well-nigh impossible, barge-loads of commodities such as hides, furs, timber, and ginseng were floated down the Kentucky River from the eastern part of the state to market here, hundreds of miles away.

It's a hard city to get a handle on. Louisville looks northern, its old streets lined with bulky brown brick buildings, but it talks Dixie. And nothing could be more southern than the Kentucky Derby, which sends the whole city into an orgy of mint juleps and parasols every May. But the names in the phone book are downright Midwestern—columns of good German Schmidts and Schumachers. Even the locals disagree about how to pronounce the city's name. A lot of them mash it down into two vague syllables—LUHvle. A few, including Steve Goodman, come out with a more enthusiastic LOO-ey-vill.

So, who is the S. in S. Goodman and Sons? I ask, after the Chinese ladies have bustled off down Market Street.

"It was my great-grandfather," says Steve. "His name was Simon Goodman. He had four sons, and then the four sons had children, and then my father was from one of the four sons, and he was the third generation. And then I'm the fourth. And my son's the fifth generation. It's kind of been traditional that you have a son whose name starts with an S. Like my great-grandfather's name was Simon, and my grandfather's name was

Simon, and my father who was third generation was Simon, and my name is Stephen—starts with an S—and my son's name is Seth. Over the years we had different types of business. With the cowhides and the wool we had crews, but now with the root business we're down to family—my wife, and my son, and myself."

Seth, with his shaved head and teenager's build, doesn't look old enough to have been working in the business for ten years, but it turns out he's actually twenty-nine, and baby Shayna is his daughter. After a couple of years at the University of Wisconsin, where both his sisters had gone, he decided to come home and learn the root trade. He works in the warehouse and "tells me what to say on the phone," Steve jokes.

On the wall, I notice a neat sign, printed out with that iMac. It reads:

> *Notice: We will be paying full price for top quality wild ginseng only! Top quality wild ginseng does not include cultivated, woods grown, dirty, scrubbed, cut up, broken, or small root. Burnt root has no value.*

I ask Linda about this. "Sometimes when people are starting out, they just don't know any better," she tells me. "They just don't care for it properly, and they dry it on a piece of metal or in an oven, and it's no good."

The sign makes me think of the diggers and growers I've heard complain about the dealers, calling them swindlers who try to hide the real value of the root. "Do people often argue about the value when they come in to sell?" I ask Steve.

Diggers, he feels, have little sense of the quality of their roots. "If you see one piece of ginseng or a few roots, and that's all you see, you don't ever have an idea. But when you see hundreds and thousands of pounds a year, you pretty much know what ginseng is. It's the same thing with furs, the same principle. Many years ago, we used to buy raw furs, raccoons and things like that, and some guy came in one time and we were grading his furs, and he said, 'I know this one's good quality, it should be better than that, it should be better than that.' And I tried to explain to him, 'Sir, that's all the value that's worth.' He said, 'Aw, I've been trapping for years, and I *know* that's a better quality,' and I said, 'That's all we can do and you can do what you want,' so he left. And I was telling my dad, I did all I could to try to satisfy him and explain it to him, but kept saying it was a different quality. And my dad said, 'Well, he's a trapper, he's maybe trapped for five or ten years. Maybe he handles twenty or thirty skins a year. We've been here for a long time, and we handle tens of thousands of skins each year, and you know the difference between a good one and a bad one.' And that's the same thing with the ginseng. When you handle thousands of pounds of ginseng, after a while you know what ginseng is and what it isn't."

He himself started working there as a teenager—his first job was folding the bags that wool fleeces arrived in. He later learned to grade wool, and then furs, but his plan was to become a teacher. He graduated from college and was doing substitute teaching when he received his draft notice. It was 1967, and he and Linda had been planning a June wedding. Instead, they were married a week later ("There was an ice storm, so they called school off, so he didn't even miss a day of work," Linda tells me).

Goodman did his military service in South Korea, but surprisingly, a single cup of ginseng tea in a coffee shop was his only contact with the plant there. Finally, he came back to Louisville to begin his career teaching high school social studies—only to realize his heart wasn't in it. Then his father's fur grader announced that he was retiring. Steve joined the business in 1969 and hasn't looked back.

The company I'm visiting, though, has changed greatly from the one that he joined as a young man. For a time, they had a subsidiary in St. Louis, the F. C. Taylor Fur Company, a receiving company for furs trapped all over the United States. That business was later moved to Louisville and consolidated.

"We operated the fur business until 1992," he tells me. "Furs are not a popular thing today, but at one time the United States was a rural country, every young man trapped to supplement his income. And at one time, St. Louis was the hub of the fur industry. And there were probably forty fur houses in St. Louis at the turn of the century. When we left, there were two left. Now there aren't any that I know of, though there's a few people in Ohio that still buy furs. Over the years, various industries declined and were phased out. Our focus now for the last ten years has been the roots and herbs, with ginseng and goldenseal being the two most prominent." I'm surprised to hear that they actually handle much more goldenseal, which grows in the same places as ginseng but is less valuable—and less romantic.

People occasionally come in to buy ginseng root from them, like the two Chinese women, but most people would rather just go to the drugstore and buy it in capsules. "Americans have a McDonald's philosophy," Steve complains. "If they can't go through the window, pick it up, and do it right then, they don't have time for it. Most of the Chinese will tell you there's several ways that they take ginseng. One way is that they'll cut up a piece of ginseng like you would a carrot and put it in a soup and simmer it. Americans don't have time for that. If you can't open up a little can of Campbell's soup, heat it up for two minutes. . . . And a lot of times the Chinese will tell you they put the ginseng in some cognac for a month or two, then drink that, just a little sip."

According to Linda, the ginseng diggers who come in to sell their harvest are generally middle-aged, and "99 percent" are male. Beyond that, she says she can't generalize. "Most of the people who have been in are, say, people who wouldn't conform to a nine-to-five job. The majority do it for extra money and love nature, love being outside. And the majority of them will be people who were fortunate enough to have somebody that taught them—either a father, or uncle, or grandfather, somebody they met along the way who could show them."

Steve adds, "We used to have lots of people who would come

in here and say, 'My father used to bring you roots and furs back in the old days. . . .' We don't have as much of that as we used to, but you still do have some. They're out squirrel hunting, and they come across it, or people that are out in the woods that live near, retired people, or you get people who are doing work, maybe construction, and they come across it through their surveying. In some areas, mainly Appalachia, they're on supplemental income, or other assistance, so ginseng is their living."

Each state has its own season for buying ginseng—in Kentucky, it starts September 1 and runs through March 31 of the next year. Most diggers, understandably, are eager to get their money and come in early in the season. September and October are when the Goodmans get most of their "door trade"—people who show up with roots to sell. A few will hold on to their ginseng until right before the holidays—apparently the tradition of digging ginseng for "Christmas money" still lives on in Kentucky. After that, business is sporadic, on through the end of March. The rest of the year, the Goodmans are occupied with the other roots and herbs they deal in.

"We handle goldenseal root, we handle goldenseal herb, which is the leaves off the goldenseal, and we handle a dozen other things—echinacea, bloodroot, cranesbill, star grub root. . . . Most everything except the ginseng, which goes to Asia, is used in the United States for the most part. I think the industry says there are two hundred salable roots and herbs from time to time in the industry, and there are probably from twelve to fifteen that are actually viable in any one year—that have some value to them. Things like slippery elm, witch hazel. There's just dozens of different roots and herbs, but none of them have the value of ginseng or goldenseal."

The other herbs are sold to nutritional supplement companies—you may have consumed some of the Goodmans' echinacea in those capsules you bought at your drugstore. But ginseng is another matter.

"What happens with the roots in this country for the most part, they're sold to Chinese agents, who represent their families.

We might call them corporations, but they're really families. There are four or five major families that deal with ginseng, and they have their agents, whether they're family agents or representatives for them in the United States. You'll sell the ginseng for the most part to the Chinese, who'll export it overseas. They initially come to you. Once they know who you are and the type of root that you handle, after that it's mostly just shipping it to them. When you have enough that it's worth shipping, or when the market conditions are right, or when they're interested and they want to call you for it, then you put an order together and ship it to them. There's some in New York, in California, and in Wisconsin."

Because trade in wild American ginseng is regulated under international law, there is plenty of paperwork that accompanies each sale. In Kentucky, when diggers come in to sell wild ginseng, they complete a form listing their name and address, and the county it was dug in, along with the amount that they have. A state official must come out to certify the ginseng before it can be shipped. They issue a certificate stating the amount of root and the year of harvest, which is shipped with the roots. Exporters must have certificates for all the roots in a shipment.

So, what's the market price for wild ginseng now?

"Ginseng right now is around $250 a pound, as a starting price. It varies anywhere from $250 to $275. As far as the start of last year, it's about the same. It changes weekly, monthly, depending on what the demand is in Hong Kong or Asia. The range in 'sang in the last couple years has been between 250 and 350. At one time, it was $520, at one time during that year, not the whole year. One year I think it was around $400. But those are the exceptions rather than the rule. Sometimes it's the fact that the weather made the crop extremely short, and there wasn't enough supply for the demand, and that's happened once or twice over the years, but in the last few years we've had a steady demand and a steady supply."

But the entire ginseng trade could end tomorrow, if the federal government changed its mind. Under international law, in order to export wild plant and animal species, government scientists must certify that the trade won't threaten the survival of the species. Since the 1970s, there have been times when it looked as though the government was going to ban the trade in its third century.

The first was when the United States signed on to a treaty called the Convention on International Trade in Endangered Species of Wild Fauna and Flora in the 1970s. As Steve remembers it, "The federal government knew very little about the ginseng business, and the following year they had a meeting in Kentucky, and all the dealers went to meet with them. And they said, 'We don't think we can allow you to dig ginseng because it only grows in four or five counties, and it's endangered.' And all the Kentucky dealers were sitting there saying, 'No, that's not right, there are 120 counties in Kentucky, and here's the counties we buy from.' And everybody wrote down the counties, and before it was over, there were 116 of the 120 counties that they'd bought ginseng from." Obviously, ginseng was more common than anyone in Washington had imagined, and export was allowed to continue.

It happened again in 1999. Goodman recalls, "A few years back, one of the federal authorities announced that they were going to restrict the shipment of ginseng, and that scared everybody." The government put rules into place specifying that only mature plants five years of age or older could be dug for export—the first-ever restriction imposed. "And the whole industry began to meet, the dealers came from all over and we met in Kentucky. And we hired an attorney so that we'd have someone to be a spokesman for us. And that evolved, and finally we were able to resolve that issue, that they weren't going to stop the export of ginseng."

I wonder, though, whether wild ginseng is truly safe. Several people now have told me that the number of ginseng roots per pound has been steadily rising over the years, all across the natural range of wild ginseng. This means that the plants being dug are smaller and younger, a real threat to a species that takes three or more years just to produce its first seeds. Doesn't this bode ill for the future of a ginseng industry?

Steve isn't alarmed. "Let's assume that there's some basis to that," he says. "What could contribute to that is the type of diggers we have. If they're just in a hurry and don't take the time to go far enough back, if you go further back in the woods you're probably going to see larger roots and they'll be less per pound. That's a hard thing to definitely say. I know the state has to do that, they count the numbers from time to time for representative samples, but I don't know how valid that is because it depends on what section you're doing of the state and the fact that— Our old-timer diggers, whether it be of ginseng or any other herbs, are fewer and farther between then they were, and the young people are in a hurry to get what's close by and not take the time to go into the woods."

"And the other side," Linda adds, "is there's so much construction going on and people are going further out into the country building. And a lot of people will bring in root and say, well I just dug this because it was there, and they were clearing the land for such-and-such."

"Like I said," Steve goes on, "we've been here for four generations. And if we don't conserve the plant, it won't be here for my son. This is not just a job with us, it's not just our livelihood. It's been who we are, for five generations."

We step out the back door to enter the warehouse, a low brick building that looks at least as ancient as the office. A dry, sweet smell drifts on the air. Rows of waist-high brown cardboard drums line the walls, and Goodman pries up their metal lids to show me the different herbs his firm deals in. Goldenseal is called "yellowroot" in the hills of Kentucky, and when he snaps a root open, I see why: the break shows a lemony color. Bloodroot is an alarming beet-red inside. For shipping, many of the herbs get pressed into two hundred–pound bales in a square blue machine, then wrapped in burlap. "If it's going to a manufacturer, we make it look a little more professional," he grins. Wheeled gray canvas hampers filled with thready snarls of roots await their turn at the baler. It amazes me to think of this anachronistic business still carrying on in the heart of Louisville, a couple blocks from their spanking-new glass convention center and the downtown office blocks.

Behind the brick building there's a smaller wooden building filled with still more barrels and hampers, with something like a giant garage door at the back. All the warehouses on the street once had a similar building, he tells me, stables for the teams of horses that hauled the goods.

Then he presses a button and the door opens onto the alley—and a sonic assault. Just across the alley, air hammers are chewing the walls off a building that's slated to be replaced by condominiums. And next door to the Goodmans a hotel is going up—they're pouring the foundations now. "On the other street, the facade of that building has a lot of architecture, and there was a discussion that went on, back and forth with the developers and the preservation groups. What they're going to save is the front facade of the building on Main Street, a little portion of it, and they're going to put it in the lobby when they build it. You won't know that any of this was ever here."

I ask how he feels about the neighborhood changing.

"It's inevitable." He shrugs. "If you look at when I first started working in the old S. Goodman and Sons, there were old hotels, all different things—they're all gone. It's progress." He doesn't sound too disturbed. "My building that I came out of goes back to 1858, and when we remodeled it, we had to go to the development commission and we had to do it a certain way to preserve it. And if you're going to preserve an old building, and keep it, and use it, that's one thing. But you don't preserve something if you're not going to use it. That's my feeling. If you're not going to use this thing, what's the sense of having this old landmark building and architecture? You can take pictures and you can look at the books if you want to see what it looked like."

I can only imagine that when your family business has made

it through 130 years you have a pretty good sense of what's actually worth worrying about.

But there are people who would like to put the Goodmans, and every other ginseng dealer, out of business tomorrow. Some environmental groups and scientists believe that American ginseng will survive as a species only if all exports are banned. Since 95 percent of the wild harvest is shipped abroad, that would mean the end of the industry.

No one disputes that there have been changes over the years in ginseng. Diggers admit they have to walk further and hunt longer to fill their sacks and pillowcases. Jo Wolf laments the loss of dozens of sites where the plant once flourished, now built over with subdivisions or trailer sites, or blasted into the sky for strip mines. Official statistics in many states record a steady decline in the average weight of ginseng roots harvested—meaning the plants are being dug up younger and younger, before they have a chance to seed the next generation.

In response to this, the Appalachian Ginseng Foundation, Syl Yunker's group, called for a moratorium on all exports of wild ginseng until research could be carried out to determine whether the wild harvest could actually be carried out in a way that wouldn't doom the species.

Wild ginseng has been disappearing for a very long time indeed.

In 1901, people said it was on the verge of extinction. Dr. John Uri Lloyd, in a paper on American ginseng presented before the American Pharmaceutical Association, said, "But as the woods have been mostly cleared off and the thickets cleaned out, this plant, which never grows in beds and is always very scattering at the best, became scarcer and scarcer, until now it is nearly in a condition of extermination."

In the 1880s, people said it had nearly vanished. Pioneer grower Val Hardacre wrote, "The forests had diminished to the point at which they could no longer supply the demand for wild ginseng, and the prices paid each year mounted higher and higher. A few men in widely scattered localities, wise in the lore of the woodlands, began to experiment in wild ginseng culture," hoping to keep the lucrative trade alive.

In 1749, people said it was almost wiped out. In that year a Swedish explorer named Peter Kalm wrote, "Many people feared lest by continuing for several successive years to collect these plants without leaving one or two in each place to propagate their species, there would soon be very few of them left, which I think is very likely to happen, for by all accounts they formerly grew in great abundance around Montreal, but at present there is not a single plant of it to be found, so effectively have they been rooted up."

Though everyone agrees that wild American ginseng has gotten harder to find over the years, the first actual attempt to protect the plant didn't come until 1975. In that year, *Panax quinquefolius* came under the protection of the Convention on International Trade in Endangered Species of Wild Fauna and Flora. When the United States and 144 other nations signed on to CITES (pronounced SIGH-tees), they agreed to ban exports of all endangered species, and to keep a watchful eye on trade in all species whose survival was threatened. American ginseng was included in Appendix II, the list of plants and animals which, though not currently endangered, "may become so unless trade in specimens of such species is subject to strict regulation in order to avoid utilization incompatible with their survival."

Species on the Appendix II watch list can be exported only if that country's government has determined that export doesn't threaten the survival of the species, and that its harvest was conducted in accordance with the country's wildlife protection laws. In the United States, regulation of ginseng is carried out by the U.S. Fish and Wildlife Service (USFWS). Every year, its Office of Scientific Authority (OSA) issues a report called a "finding" which determines whether the international trade in wild ginseng can continue.

The finding for the 2003–2004 season runs to dozens of pages, summarized at the beginning in a single pithy paragraph:

Please be advised that, based on our analysis of available information, we find that the export of wild American ginseng roots of 5 years of age or older harvested during the 2003–2004 seasons in the following States will not be detrimental to the survival of the species: Alabama, Arkansas, Georgia, Illinois, Indiana, Iowa, Kentucky, Maryland, Minnesota, Missouri, New York, North Carolina, Ohio, Pennsylvania, Tennessee, Vermont, Virginia, West Virginia, and Wisconsin. One little change in the wording of this sentence, and a multi-million-dollar industry would end.

For the Appalachian Ginseng Foundation, the situation was clear-cut: "The Office of Scientific Authority must establish a moratorium on wild ginseng exports." In a technical paper produced in 2000, the AGF accused the Fish and Wildlife Service of basing their analysis on insufficient data, and bowing to pressure from the industry. "In the absence of good science," they charge, "the OSA has been and will be essentially 'flying blind' when making its findings." They argue that the only way to accurately assess the status of wild ginseng is to look at the dynamics of populations over the whole of the plant's range, something that has not been done for lack of funding. Until there is a scientifically valid analysis of the situation of wild ginseng, allowing exports of the roots to continue could bring the extinction of the species in the wild.

But the trade has gone on since that manifesto was published. State governments have continued to do their counts, numbers have been sent on to Washington, and the U.S. Fish and Wildlife Service has continued to issue annual findings stating that populations are healthy enough not to be threatened by harvest. How many wild ginseng plants are really out there? It's somewhat akin to asking, "How long is a string?"

The Fish and Wildlife Service works together with the state governments to monitor the ginseng trade. Each state is required to have regulations to ensure that harvest won't harm the wild ginseng population. Ginseng dealers must register with each state in which they buy and sell roots, and must report every transaction to the state. The states must inspect the roots and certify that they were legally harvested, and compile all the data to pass along to the USFWS. As an informational handout from Tennessee's Department of Environment and Conservation warns, "Without strong cooperation from the ginseng dealers of the state on providing the data on the quantity of wild ginseng harvested and the counties from which it was collected, the OSA may rule against export of Tennessee ginseng for the upcoming year." It seems like a nerve-wracking way to run an industry.

The nondescript office building in Frankfort, Kentucky, feels light years removed from the muddy forest world of wild ginseng. But the state's venerable capital city, lined with dignified old brick buildings from the 1790s, actually has a centuries-old link to ginseng, since the days when it was a destination for flatboats laden with the root that floated down the twisting forks of the Kentucky River. Frankfort still plays a key role in the state's ginseng trade.

I'm lucky Chris Kring is actually in his office at the Kentucky Department of Agriculture today, and not out driving the state's most obscure back roads. Kring is a Kentuckian born and bred, from the peaceful Bluegrass town of Versailles (that's vur-SALES, here). In his work as the state's official Ginseng Certifier, he is responsible for inspecting and licensing the state's seventy-four dealers of wild ginseng—people like Steve Goodman. Driving thirty thousand miles a year around a smallish state, Kring has worn out a number of state vehicles, and one of his own. "I've seen every end of the state in twenty-seven years, every county," he tells me. "Even Inez. And you don't go to Inez unless you're going to Inez, you don't go through there to get to anywhere!"

It's been an adventure. He still remembers his first trip to the mountain town of Pikeville, years back: "It was a state highway, then a county road, then I was on gravel, then I started going back up this hollow, a typical—stereotypical!—scene of junk cars, washing machines on the front porch, people coming out of their houses . . . and I was in a state car with a big seal on the side. I saw a couple of shotguns, and I thought, what in the *world* have I gotten into. . . . I got to the last house, and this old guy was standing out on the front porch, and he was laughing. And he said, 'Come on in here, boy!' He knew exactly. They were his family, all on through there. I stayed a couple of hours."

Of the ginseng dealers he visits and certifies, few are large businesses like the Goodmans'. Most dealers work other jobs, and only a handful are full-time, year-round dealers who buy many different roots. In the past, many of them were fur dealers, but the fur trade has just about vanished in Kentucky. Nowadays, a lot of ginseng is bought in mom-and-pop grocery stores, or in other small shops—in one town, it's the local furniture store. In another county, a school administrator took over her father's ginseng business when he died, and ran it as a sideline for twenty years until she retired. Other states are similar—the list of registered ginseng dealers issued by the State of Tennessee tells a tale of its own. There's Smoky Mountain Herbs and Medicals, but there's also Bristol Scrap Metal, Jack Seals Grocery, and D & L Pawn Shop.

Certification is a fairly simple matter. Dealers keep records of their purchases, and turn over to the state lists of the diggers' names, the number of pounds, and the county where the root was dug. Kring then issues the certificate that goes with the ginseng so that the roots can be legally exported. "I do root counts sometimes, and sometimes I have them weigh it all out. It doesn't matter if they've got five hundred pounds of ginseng there, if they've only got papers for two hundred pounds, that's all the certificate they're going to get."

Kring is the first to admit the inherent conflict in his job: he's supposed to be protecting the species—and at the same time, promoting it. The information collected by Jo Wolf about the populations of wild ginseng around the state is transmitted to him, and he compiles the annual statistics that are sent to the U.S. Fish and Wildlife Service. Along with the reports submitted by the other states, this is what ultimately determines whether the trade in ginseng can be allowed to continue. On the other hand, he's also supposed to be helping to sell more Kentucky ginseng, a part of his job that he devotes far less attention to. If ever there was a product that sold itself, it's ginseng.

"Ideally, in terms of protection, this should probably be someplace else besides the Department of Agriculture," he says. "But nobody else wants it, so we've got it. We're showing that ginseng is still here, year to year, and I'm showing that it's also a significant economic factor, especially in some of our hardest-hit counties. So yes, we want to see it continue."

But does wild ginseng have a future?

"It will be all right," he says firmly. In 2003, the last year for which he has complete statistics, over 22,000 pounds of root were dug, in 106 of the state's 120 counties. He admits that this figure has declined slightly in recent years, but says this is not significant. And other states have not shown large declines—if that were the case, it could indicate that their root was being sold in Kentucky instead. Eleven tons of wild ginseng—the roots from something like six million individual plants—were found and dug in this one state, in one year. "Overall, it's there, it's still reported."

A complicating factor is that it can be difficult to determine what, exactly, is wild ginseng. For one thing, a lot of people (Syl Yunker among them) have put a lot of effort into growing roots that look exactly like they were dug in the wild. It's not unheard of for diggers to bulk up their harvest with woods-grown roots, thus adding them to the official state statistics for wild ginseng. As far as Kring is concerned, any root that the Chinese will pay the wild price for is wild. But there are many other gray areas. What about the landowner who tosses seeds off his back porch every year and lets them fend for themselves? The seed may have come from a farm in Wisconsin, from a different gene pool entirely, but the plant will be reckoned as wild Kentucky ginseng.

"With all the cultivated seed that's been bought and sold," Kring says, "and land that's cleared and then reforested—that ginseng's there, but where did the seed come from? At what point is that a wild plant now? We know that what was once pasture and cropland is now twenty-year-old forest, and ginseng is growing there and being harvested and sold as wild. So is Eastern ginseng in the wild truly endangered?"

I ask his opinion of the ban on wild ginseng export which some have argued for.

He shakes his head knowingly. "That won't do anything except drive it underground. It'll become a black market. It's real easy to get something out of the country. It's hard to get something *in,* but you can go to Newport News, Virginia, rent you one of those big containers, put whatever you want into it, and chances are real slim that anyone will look at it. They put it on the boat and hey, it's leaving. You can go to any seaport and ship it anywhere you want, uninspected. All you're gonna do is drive ginseng underground and make everybody criminals. Because it won't stop it."

The situation is complex, to say the least, but certain facts are not disputed. Wild ginseng is getting harder to find. The harvested roots are smaller and smaller. The plant's wooded home is being devoured by everything from coal mines to golf courses to trailer parks. Can we still, responsibly, allow the roots to be shipped off by the ton to Ko Shing Street in Hong Kong?

Hoping for an answer, I call Dr. Jim McGraw, a conservation biologist at West Virginia University whose name has come up again and again in my conversations with ginseng people. How, I ask him, can I know who to believe?

He gives me a deceptively simple reply. "What I would ask is, how much of what anyone is saying is based on science? And how much could be based on their vested interest in having it come out a certain way—having their opinion believed as truth. Because a lot of these people are in it for money. And then you know how much to believe it. There's so much lore, so much made-up stuff because of the stories people like to tell."

It's not true, for instance, that "ginseng doesn't come up every year," as a half-dozen people have sworn to me. The plant is actually a favorite food of white-tailed deer, who will munch it off down to the ground—though the root remains, and will send up a new plant again. But the only way to know this is to go out and count the populations very early in the spring.

He also debunks the idea that there is no such thing as wild ginseng any more because all of it is cultivated ginseng left over from when there were more ginseng growers everywhere. "We *know* that's not true, based upon the genetics," he says. "We've done the genetics that shows a clear distinction between the wild and cultivated. A student in Maryland just did this, she went to a wild population and discovered a couple of cultivated genotypes mixed in. She was able to find genetic markers that clearly showed that these seeds came from Wisconsin, while these other seeds are local. So, that's the kind of thing I'm talking about. It is true that there was more ginseng cultivation in the past, but that doesn't mean that people were spreading seeds through hollows that were fifty miles from anywhere, they weren't *doing* that. At least not in Appalachia! So these claims that we don't need to worry about cultivated plants because it's all cultivated—that's ridiculous. It's not true."

But there are huge gaps in what science knows about ginseng—a lot of it simply because there aren't enough people out in the field checking up on wild plants. And it seems a lot of current policy is based on wrong information.

Take, for instance, the starting dates of the legal ginseng harvest. Because ginseng is such a slow-growing plant, which reaches peak seed production only in its fifth year of growth, it's crucial that people dig the plants only after the berries are ripe and ready to go into the ground. Diggers are in a hurry to get their money, but the states have established a starting date for the season when the berries are red and ripe—a date which ranges from August 15 on through the fall.

McGraw and thirteen other researchers set out to determine the dates when ginseng berries ripened at thirty-nine separate sites, from Maine to Missouri to Virginia. Unfortunately, their careful surveys found the date could not be predicted by geographical location alone. This means that states had simply been guessing blindly about when to let the harvest begin.

Another complicating factor, according to McGraw, is that ginseng seeds actually have a higher rate of germination when they are carefully planted under an inch of soil by a human than when nature lets them fall to the ground to be eaten by animals. Once the berries are ripe, it's actually better for the species to have the plant dug by a conscientious 'sanger who will plant the seeds than to let it go through its natural cycle.

But given all these facts, how could the team of researchers recommend any specific digging season? Instead, they ended up talking about trade-offs, the pros and cons of different dates. An August 15 start was long before the berries were ready, but as it was already the law in the four largest ginseng-producing states, retaining it would upset the smallest number of people. By September 15, most of the berries are ripe, but many have already fallen off—and the date would mean changing the regulations in nearly every state. They finally settled on September 1, a time when most berries are ripe, and those that aren't have a good chance of germinating if planted. It's also the actual starting date in nine states.

"But," he points out, "we're not policy makers, we're just laying out the science. It's the Fish and Wildlife Service, along with the state programs, that are going to have to decide what to do with this science, which they've never had before."

And the situation isn't likely to change. Though each state is supposed to have an ongoing program to monitor the status of wild ginseng, there is no uniformity. Jo Wolf, with her five-year cycle roving across Kentucky, has no counterpart in other states—and every year Kentucky has to scrape up money, a grant from somewhere, to allow her work to continue. One year the grant came through so late that she was hardly able to get out before the harvest.

"The problem is, this is an unfunded mandate from the federal government," McGraw tells me. "The Fish and Wildlife Service says, 'OK, states, you need to have a ginseng program,' so they put it somewhere—they put it in a heritage program, a wildlife program, a forestry program, they stick it somewhere.

And someone is put in charge. And then every year the FWS wants a report from them, so they provide them the most minimal data they possibly can, because no one's giving them additional money to do this. They have to produce the data that says OK, the harvest is sustainable, but no one has the personnel on the ground to actually do that. Jo Wolf is the best that any state has that I know of, in terms of actual monitoring, and Kentucky is always begging for a little bit of money from somewhere else to get that done. The state of West Virginia doesn't even have that."

Knowing everything he knows, what is his own feeling about the future of wild ginseng?

"If diggers use best harvest practices, then it's sustainable. But it could tip one way or the other." According to best estimates, there are 87 million ginseng plants in West Virginia, spread out over a huge area; every year about 4–5 percent of them are dug. Whether that figure is sustainable over the long term depends on exactly how it's done. If diggers take only mature plants and plant the seeds correctly, the success rate is high. But how many diggers follow these practices is anybody's guess, and nearly impossible to measure. What people say they do, and what they actually do, can be two different things. And few diggers are likely to let a scientist follow them around. There's really no way to know what is going on along the mountainsides. McGraw can only hope for the best, and keep on trying to supply the missing facts that policy makers need.

For McGraw, ginseng's true worth goes beyond the formidable economic statistics, and even beyond the plant's environmental significance. "There's a lot of cultural value to this ginseng activity that goes on. We estimated that in West Virginia something like fifteen thousand individuals were going out every fall, taking hikes in the woods. People do this often as families, often across generations, and what are they doing? They're learning to identify a plant, they're learning about the value of the forest from a different point of view—it's not the timber value, it's not just pretty trees, it's that there are a whole host of plants in the forest that are actually interesting and valuable in themselves."

To become a successful sanger takes years, during which people acquire a detailed understanding of all the plants to be found in the forest, which ones grow together, and where they grow, and why, and how.

"They become good natural historians, self-trained, bright, natural history experts. And that's tremendously valuable—you can't put a price tag on that. You have fifteen thousand people going out and doing that every year, and the tie that's created between the people and the land is really valuable in terms of preservation of the environment for the long term. These people have a buy-in. They *care.*"

So, in the end, everyone may be right. Wild ginseng is in trouble, *and* stricter regulation is needed, *and* we don't know enough, *and* the harvest can be not only sustainable but be a good thing for the species. I've noticed that ginseng "experts" often give little attention to the big picture, getting wrapped up instead in nailing down the fine points of licenses or genetics or mulch. People have reams of documents to prove they're right and pointed arguments to show their opponents are all wrong.

What they all fail to notice, though, is that ultimately they're all on the same side. Poachers, diggers, dealers, bureaucrats, conservation biologists, the men behind all the desks in Ko Shing Street in Hong Kong—no one wants to see ginseng go extinct.

SEVEN

UNDER THE MICROSCOPE

They're almost beautiful as they slowly drift along, these dazzling, transparent shapes beneath the lens of the microscope. Some glitter hard-edged as diamonds, others are softly irregular, and the round ones flash like clear sequins. I shiver in spite of myself. In Dr. Laura Murphy's lab, in Carbondale, Illinois, I feel like I'm staring down at death.

"All of these are breast cancer cells?" I ask her.

"Yeah. . . ." Murphy's bubbly voice has turned serious, almost wistful. "They're in different stages of division," she explains. "They're rapidly, rapidly dividing . . . and they don't die. That's what makes them cancer cells. They just keep growing, growing, growing. They grow and fill this flask up, and we have to divide them and put them in another flask. They actually form little tumors in there, if you let them keep growing."

"And that's what the ginseng slows down?"

"Yes," she says simply. "That's right."

In her research at the Medical School of Southern Illinois University, Murphy has discovered that an extract of American ginseng, prepared with roots from Syl Yunker's farm, can stop the growth of breast cancer cells. And her project is hardly the

---

Previous page: In the system of traditional Chinese medicine, ginseng is the most highly prized of herbs.

obsession of a lone health-food crank—it's funded by the U.S. Department of Defense.

"Do you ever think about what these really are?" I ask, as she removes the flat, transparent flask from under the lens.

She nods. "These are real cells, from real people. Who *died.* I can't forget that." I must look as unnerved as I feel, because she reassures me that they can't do any harm if they spill from their container. I won't get cancer from them. But then she admits, "The first time I looked at these cells, I put on a mask, put on gloves. . . ."

Laura Murphy is slender and animated, with short, sandy hair and clear blue eyes. On a raw December day in the Midwest, she's bouncing around outside in a sweatshirt and Tevas—sheer enthusiasm seems to be keeping her warm.

The physiology lab where she carries out her research has the air of a home-away-from-home, a long pastel corridor lined with offices and workrooms where people obviously spend way too much time. *The whippings will continue until acceptable results are obtained,* jokes a sign on one lab door. On another, a poster says *The universe is filled with magical things patiently waiting for our wits to grow sharper.* Though work around the lab is slowing down for the Christmas holidays, it never really stops. A graduate student is sitting and writing up notes at a desk, and one unlucky technician will spend the vacation tending to the cancer cells.

"You have to feed them every other day," Murphy explains, "with that yellow media in there, and you have to really monitor them because you don't want them to get too thick, you have to split them into new flasks. . . . They require a lot of care. So over Christmas we've frozen everything down in this liquid nitrogen, and we'll thaw them out and bring them back up after New Year's."

Using ginseng to treat cancer is not a new idea. Nearly a hundred years ago, stories were floating around about how ginseng, the "mystery plant" so treasured by "Orientals," could heal people of tumors. In the February 1916 issue of the *Ginseng Journal and Goldenseal Bulletin,* an article entitled "Ginseng Cancer Cure" by J. N. Croddy reports on a man who had suddenly developed a huge sore in his mouth that began growing at a frightening rate. After several months of pain and anxiety, and unsuccessful treatment by his physician, the man began using ginseng. Croddy quotes the unnamed patient, whose father and brother had died of cancer: "I was beginning to have visions of a horrible and lingering death, but the beneficial effects of the ginseng were apparent at once, and very much to my surprise and delight the cancer rapidly healed and within a short space of time had disappeared altogether."

Croddy mentions two other cases of what appear to be an oral cancer that were healed by applying pulped ginseng root as a poultice and eating ginseng roots. He admits that reports like these don't constitute a proof, but he feels confident that science will soon confirm these observations. "Ginseng," he says, "is one of the most wonderful medicinal plants in the universe today, and while I do not lay any claims to being a prophet, I am going to make the prediction that the real medicinal virtues of ginseng will be discovered and given proper recognition by the medical fraternity in the not far distant future."

That's precisely what Dr. Murphy is hoping, almost ninety years later.

Over eggs and pancakes at the local Denny's, Murphy tells me about her recent work. She's just back from giving a paper at an international ginseng conference, in Australia, of all places. It brought her a totally different perspective from that of the ginseng community in the United States, which is a somewhat inbred group that shares all the same problems: "Folks here are dealing with people stealing their ginseng," she tells me, "and also the fungus and rotting and issues that come up in trying to get the root out of the ground. . . . There, it's more predation by strange marsupials!" Though it seems unlikely, ginseng cultivation has been fairly successful in Australia.

Because she was one of the first speakers on the program, people kept coming up to her throughout the conference asking for more information about her work on ginseng and cancer. She was especially surprised at the warm reception she got from the farmers, who she'd figured would be concerned only about advances in growing techniques. "I think that folks are really interested in the justification for why they're growing their ginseng, and so when they hear that ginseng has these potential anti-cancer properties, and that these are reports that are coming from scientists in North America, who have funded research that's actually looking into this area, it gives them a really positive attitude about what they're doing. So I got a lot of pats on the back from growers and farmers happy to hear that ginseng had this potential, which may really increase its marketability."

The tale Murphy tells me about how she got involved with ginseng sounds pretty darned far-fetched. As an associate professor in SIU's Department of Physiology, her chief responsibility was, and is, teaching courses to medical students on the human reproductive system. For years, her research focused on marijuana and its effect on reproduction. (Among other things, she discovered that when you give marijuana to pregnant rats, their male offspring aren't interested in sex.) Then, one year, she found out that she was going to have some high school students as lab assistants for the summer, under a university program. Her latest marijuana project was finished, and she suddenly found herself with a whole crew of enthusiastic teenagers on her hands.

"These were kids who were mostly interested in being physicians one day," she says, "and they had an opportunity to work in a laboratory and get paid for it. It was a program designed to widen their horizons, and show what careers were available for people who were interested in science, so this group was seeing what it was like to be a research scientist over the summer." The problem was to come up with something useful that they could actually do.

"I was working with marijuana," she recalls, "but my brother and his father-in-law in Tennessee had just decided to plant ginseng. And my brother called me and said, 'I've met this guy named Scott Persons, and he's sending us some seed, and we're going to plant some ginseng—tell me, what's in it? What's it do? Can you just look it up and give me some information on what it's good for?' And so I did the lit review, and I'd just finished the marijuana and libido study, and looked up what the ginseng does, and one of the first things they mentioned was it's a tonic, and it's been shown effective in the treatment of impotence, and stimulates libido. I thought, oh, really! Stimulates libido, interesting! Because marijuana has the opposite effect." Contrary to popular belief, lab animals given marijuana lose interest in sex.

"I looked in the Sigma Chemical Company catalog and saw American ginseng, and I thought, you know, this might be a little project we could do. Sex and drugs, what a great project for high school students to get them interested in science!" The plan was to give ginseng to the rats and time how soon they mated afterward, then compare this with a control group of untreated rats to see what effect ginseng had on the sex drive.

"And I have to admit," she says, "I was just as guilty as the scientists that are out there, regarding, was ginseng going to have an effect? Probably *not,*" she says, in an exaggerated tone of dismissal. "I just thought, we'll see what happens. And it'll teach them how to deal with negative results, because that's part of science." Every researcher knows that discovering what doesn't work is just as important for science as discovering what does—though that doesn't mean it's just as gratifying.

So she designed her study. The rats were divided into two groups, one of which received powdered ginseng mixed with sesame oil, and a control group, which got just the sesame oil. The students came in every morning to feed the animals with a blunt-tipped syringe. The rats loved the ginseng's flavor, and lined up to get their dose when they saw the students arrive. "It was kind of a neat study," Murphy says. "It gave [the students] a lot of positive feelings about the research experience, because they really enjoyed it.

"But what was so *astounding,*" she says, her eyes getting wider, "was actually looking at the results. It was done blind. I treated the animals that day, so no one knew what animals were treated with what. The students watched the rats to time how quickly they mounted the female rats and ejaculated, because those were the two indices we were looking at.

"And it was—night and day," she says, simply. "The animals that had gotten ginseng had a much higher libido and performance rate than the control animals. I mean, cut and dried. I was sitting there astounded because I had never seen or heard of a compound that had that kind of an effect. It was that profound. Even prescription drugs out there for libido aren't as strong as the effect that we saw for ginseng."

Though ginseng is viewed in North America as an "alternative" remedy more suitable for the shelves of health food stores, it does have a history in Western medicine. When it was first brought from Asia to Europe in the seventeenth century, physicians experimented with it as a treatment for wasting illnesses, and found that it had benefits for patients with respiratory problems. Several of them reported their successes to the Royal Society of London, the most important scientific organization of the day.

With the discovery of American ginseng, investigations proceeded on both sides of the Atlantic. In 1728, a scientist and surveyor in Virginia named William Byrd reported: "Its vertues are, that it gives an uncommon Warmth and Vigour to the Blood, and frisks the spirit, beyond any other cordial. It cheers the Heart even of a Man that has a bad Wife, and makes him look down with great Composure on the crosses of the World." Any benefits for women burdened with bad husbands were not recorded.

By the nineteenth century, American ginseng was stocked in pharmacies and listed in the U.S. Pharmacopoeia, the official compendium of recognized medicinal substances, though with a dismissive tone: "The extraordinary medical virtues formerly ascribed to ginseng had no other existence than in the imaginations of the Chinese," states the 1876 edition. "It is little more than a demulcent, and in this country is not employed as a medicine. Some persons, however, are in the habit of chewing it, having acquired a relish for its taste; and it is chiefly to supply the wants of these that it is kept in the shops."

But there were other currents of medical thought in the United States. The 1909 edition of *King's American Dispensatory,* a pharmacy reference used by physicians of the Eclectic School who incorporated herbs and homeopathy into their practice, contains a lengthy write-up of ginseng. In its discussion of the root's properties and usage, it says: "A mild tonic and stimulant. Useful in *loss of appetite, slight nervous debility,* and *weak stomach.* Continued for some length of time, for its temporary administration gives but little benefit, it is a very important remedy in *nervous dyspepsia,* and in *mental exhaustion from overwork.* It gives fairly good results in *nervous prostration,* and in *cerebral anemia.* By some, it is considered useful in *asthma gravel, paralysis,* to invigorate the virile powers, etc. It gives fairly good results in atonic laryngitis, bronchitis, and some relief, in *phthisis,* being a secondary remedy for these complaints."

Publications for eclectic physicians contained occasional debate as to whether the Chinese knew something that we didn't about the plant. Wrote one editor, "If any reader . . . has studied this remedy, or will study it, and will report to us his conclusions and experiences, we will be under lasting obligations to him. We have always thought that an article that is of so much prominence in the commerce of a great nation, was certainly of some value, and that its true worth may be wholly unknown to us."

And more tantalizingly: "Aphrodisiac powers are ascribed to it and by some it is thought that to obtain this effect is the chief end of its free and frequent administration and use among the Chinese. We can neither affirm nor deny, as we have not used it sufficiently to express positively our opinion. We invite the co-operation of Journal readers. Try it and report." Alas, no reports followed.

What about the scientific evidence? Though modern scientists have been investigating ginseng and its effects for decades, these earlier studies are usually dismissed as being of dubious value. Researchers in the Soviet Union, China, Bulgaria, and Korea have done extensive work, but have often ignored international research standards. Their studies have been published in obscure journals and never translated into English, and are written in an overblown style that alienates Western scientists. Their methodology and statistical analysis are often seen as dubious at best.

Ginseng research in the U.S.S.R. began in the late 1940s, when the northern half of the Korean peninsula was coming into the communist sphere of influence—and with the land, its most valuable crop. Soviet researchers were far less interested in the healing powers of Asian ginseng than they were in its stimulant properties: could it be used to get more work out of the toiling classes?

In 1948, a pharmacologist named Itskovity Brekhman at a research institute in eastern Siberia began the first scientific attempt to investigate the tonic powers of ginseng. For centuries, traditional Chinese medicine had insisted that ginseng strengthened all systems of the body by increasing the flow of *chi.* But how on earth could a claim like that be cast in measurable, scientific terms? Brekhman decided to look at physical stamina under stress as a way of quantifying that intangible sense of vitality. He sent a hundred soldiers to run a three kilometer race, after giving half of them doses of a liquid containing ginseng, and the other half a similar-tasting placebo. The soldiers who consumed ginseng ran the course an average of 53 seconds faster than the control group.

Brekhman's experiment touched off a flurry of research. Telegraphers in the Soviet Union were given ginseng and tested on their ability to send complex messages. It was found that they transmitted messages only slightly faster than the control group, but made far fewer mistakes. In Sweden, subjects were tested on their ability to follow complicated mazes and solve puzzles. Those who received ginseng were more accurate. In Germany, subjects pressed a button to measure reaction time, and those given ginseng were faster. In London, nurses on night shift who were given ginseng reported feeling more alert on duty, and did better on tests of speed and coordination.

Meanwhile, Brekhman had moved on to testing ginseng's effects on stamina in mice, reversing the usual order of scientific experimentation. Mice were dropped into water and made to swim until they were completely exhausted. Then they were removed from the water and allowed to rest before being dropped in again. Brekhman found that injecting the mice with ginseng doubled the length of time that the already-exhausted mice could swim. Later, he tested the ability of mice to climb a moving rope, again obtaining positive results with ginseng.

Soviet scientists tormented thousands of mice in ingeniously awful ways, all in attempts to measure the powers of ginseng. The animals were blasted with x-rays, placed in freezing environments, dosed with drugs and poisons, spun in centrifuges. They were given artificially induced diseases, from malaria to cancer. Findings seemed to indicate that ginseng increased the animals' resistance to stress of all forms. But all this research was viewed in Western countries with great suspicion, rooted in both scientific and ideological grounds.

Some of the Soviet studies have defects obvious even to the non-scientist. For example, in one, a thirty-seven-year-old male pianist suffering from insomnia and fatigue was treated with ginseng for six months. At the end of that period, it was discovered that he received 12 percent more applause at his performances. Studies investigated the effect of ginseng on everything from baldness to radiation sickness, many of them without the benefit of control groups or any accurate way to measure the results.

In a 1989 book entitled *Ginseng: A Concise Handbook,* James A. Duke, at that time head of the USDA's Germplasm Laboratory and himself a ginseng grower, gave a cranky critique of ginseng research, often comparing the herb to carrots. In response to the "swimming mouse" studies, he wrote, "I think

I could swim longer had I eaten 100 mg of carrot (or tomato, or dandelion, or ginseng) per kilogram of bodyweight than if given only a salt solution."

He goes on to say, "In one account I read, mice had been given what I calculate to be the equivalent for a human of seven pounds of ginseng. Frankly, I would be afraid to eat seven pounds of ginseng (or ginseng extract) a day even if I could afford it. I would also be afraid to eat seven pounds of carrots, but at least I could afford it."

"In sum," he concludes, "I am convinced that a carrot a day is good for you and that ginseng is expensive."

Scott Persons, in his classic grower's manual *American Ginseng: Green Gold,* refuses even to discuss ginseng's health effects (or lack of them). He writes, "I originally intended to lay out that evidence at length, and to make a strong case for ginseng's medicinal powers. After reading everything I could lay my hands on (written in English) and talking with experts, I find that I cannot make the case I wanted to. . . . Much of the research is poorly controlled, unsuccessfully replicated, and/or undertaken by parties with a vested economic interest in promoting ginseng."

It's difficult to see how that last problem can be avoided. Not surprisingly, one of the main centers of medical research on ginseng has been South Korea—it seems natural that the country would want to investigate the use and effectiveness of its most distinctive crop. However, this raises questions of bias—particularly when you look at the staggering range of claims Korean researchers made for their country's ginseng.

Ginseng, they assert, stabilizes the blood sugar levels of diabetes patients. It raises low blood pressure, and lowers high blood pressure. It normalizes the liver enzymes of hepatitis patients, alleviates symptoms of menopause, relieves constipation, prevents formation of excessively acidic stomach secretions, and helps prevent breast inflammation in nursing mothers.

One Korean experiment found that ginseng sped up the metabolism of alcohol. Male volunteers had their blood alcohol levels tested after consuming the equivalent of three to four strong cocktails over a period of forty-five minutes. A week later, the process was repeated, but the second time their drinks had ginseng extract added. When the blood alcohol level of each subject was compared, most of the subjects had a blood alcohol level 30–50 percent lower when they consumed ginseng—pointing to a potential hangover remedy.

Other studies found that ginseng is effective in stimulating the formation of red blood cells. When fifty patients who had not responded to anti-anemia medications were given ginseng, their red blood cell count increased and their symptoms improved.

Korean researchers have also investigated the effects of ginseng on the liver. When animals were exposed to carcinogenic chemicals, those who were fed ginseng suffered less liver damage than the control group. When ginseng was fed to rats who had had half of their liver removed, it stimulated the synthesis of proteins and sped up the regeneration of the liver by 30 percent.

Ginseng was also found to reduce the harmful effects of radiation exposure, an important factor in the use of radiation therapy with cancer patients. Such treatment causes a decrease in the number of both red and white blood cells, but a study of cancer patients undergoing radiation therapy found that consuming ginseng returned these counts to normal. Mice recovered more quickly from radiation-induced damage to their bone marrow when they were given ginseng.

How can these conflicting effects be produced by a single herb? Some researchers, including British pharmacologist Stephen Fulder, argue that ginseng works to restore bodily equilibrium through its effects on stress hormones. Others, however, claim that all these effects are the result of government funding of research on the product of a government monopoly.

One type of research that is less easily dismissed is studies of hospital records carried out in Korea. Several different studies found that people who consume ginseng regularly have a lower incidence of a number of types of cancers than non-users. Even smokers have lower rates of lung cancer if they use ginseng than smokers who don't take it. The problem is that this type of study can't prove a connection because there is not way to rule out the involvement of other factors. For example, do ginseng users also drink more tea? Do they have a healthier lifestyle in general? But the evidence is intriguing.

It was these studies of hospital records that led Laura Murphy to wonder whether ginseng might have anti-cancer effects just as powerful as the effect she had seen on rats' libido. As we finish our pancakes in Denny's, she tells me how the focus of her research shifted away from licentious rats.

"In putting together that paper," she says, "I noticed that a lot of the research was going towards cancer, especially in Asia. Korea and China were looking at ginseng's anti-cancer effect. I certainly wasn't a cancer biologist, but one of the things I noticed was the cancers that were being looked at. The endocrine-dependent cancers, such as breast, prostate, and endometrial, had not been studied."

She explains to me that most cancers are caused by lifestyle factors—such as lung cancer caused by smoking, or skin cancer by too much exposure to sunlight. The endocrine-dependent cancers, on the other hand, are linked to circulating hormones—"kind of spontaneous cancers," she says. In men, elevated testosterone levels throughout their lives make the prostate gland more susceptible to developing cancer, just as women with higher estrogen levels are more prone to developing breast cancer. Though there are also genetic factors, for the most part these are spontaneous, naturally occurring cancers. She found it interesting that little work had been done on these very common forms of cancer.

"My training was actually in reproductive biology and endocrinology, so I was aware of the hormone-dependent cancers. I kind of thought it might be interesting—now that we'd done this one ginseng study, and people were saying that ginseng had anti-cancer effects, and I had shown that ginseng did have effects on libido, which had been reported anecdotally—could

American ginseng also have anti-cancer effects? All the studies that had been done were with Chinese ginseng. Nothing had been done with hormone-dependent cancers, and nothing had been done with American ginseng. So I kind of saw a niche. I thought, let's see what American ginseng does with hormone-dependent cancers."

The problem was getting the financial support to find out. Murphy was stunned by the negative reception that her first sex-and-ginseng study got. "We wrote that up, did a few extra measurements, and sent it to a reputable journal. And they couldn't find anyone willing to review it," she says flatly. In publishing new scientific findings, the crucial step is getting the paper reviewed and approved by other experts working in the same area. No one would touch her paper.

"I think primarily because it was dealing with ginseng, which is, by clinicians, and a lot of scientists, seen as—not exactly voodoo, but not evidence-based. A lot of anecdotal reports come out about ginseng, but actual evidence supporting any of these reports is seriously lacking. So ginseng has a bad reputation. I finally called the editor and said, 'Why haven't I heard anything about my article?' and she said, 'I'm having a hard time finding people that can review it.'" For one thing, the reviewers should be involved in similar research, which no one was doing, and the people they did send it to returned it and refused to review it. "I never heard of that before!" Murphy says. "It took a year to get that paper reviewed. It should only take three months, max. And it took another six months to get it published." Her study, "Effect of American Ginseng (*Panax quinquefolium*) on Male Copulatory Behavior in the Rat," finally appeared in *Physiology and Behavior* in 1998.

But how, I ask, did she ever get the Department of Defense to sink money into research on ginseng and breast cancer?

Murphy explains that the DOD had long been funding medical research, but was increasingly drawing criticism because the studies they supported were so overwhelmingly slanted toward male service members. They used only male subjects, and concentrated exclusively on male health problems. Then, about fifteen years ago, she said, a member of the Senate Appropriations Committee had a female relative diagnosed with breast cancer, and persuaded the committee to fund a DOD research program for innovative, unusual studies on the disease. Millions of dollars annually are now available to researchers through this program. And, she points out, "There are not many avenues for funding, and especially to get the kind of bucks you need to carry out these biomedical type studies."

Her study took a lot of effort to plan. "Cancer cells are very easy to get—you get them through a catalog," she tells me, matter-of-factly. "Trying to figure out what ginseng extract to use was tricky." Some researchers have used ginseng processed in water, while others use ginseng processed in alcohol. It makes a difference. Of the dozens of chemicals contained in ginseng, some will dissolve only in water, others only in alcohol. Murphy finally settled on a water extract. "I feel like that was more similar to what a person would do, making tea from ginseng root, and what was extracted would be similar to what we got in our extract. And the water extract of ginseng would be very easy to treat cells with." So she set out to see what effect the substance would have on cells in vitro—in a test tube.

First, she subjected samples of breast cancer cells, like the glittering shapes I had seen under her microscope, to different concentrations of the ginseng extract. "And the first experiment," she tells me, "it just worked like a charm, and I went WOW! It *does* have anti-cancer effects!" She found that the more ginseng was given, the less the cancer cells reproduced. The graph she shows me displays a clear-cut, plummeting line.

Next, she did studies on female nude mice, a special species that develops tumors when injected with human breast cancer cells. One group received ginseng extract in their drinking water, while a control group did not. Not only did the ginseng slow the growth of the tumors, but the tumors were different in appearance, and the animals who received ginseng did not develop tumors in other parts of their bodies. Furthermore, the tumors didn't kill the mice—while the control group were dying.

"I just kind of went on from there," she says. "I said, ooo, this is neat, let's see if this does anything to prostate cancer cells. And it had the same effect on prostate cancer cells, and I went wow, this is REALLY neat! So I just kept diving into that, looking at different types of cells. At that time, I reported the data at a meeting in New York, at a ginseng conference. The next thing I know all the ginseng farmers started sending me samples, and they wanted me to see if their ginseng had anti-cancer effects, so I started running all these bioassays on all these guys' ginseng, looking at the effects on cancer cell proliferation."

Her research has found that although ginseng does not kill breast cancer cells in the concentrations she tested, it acts synergistically with chemotherapy drugs to intensify their effects.

"When we give it to the mice it doesn't prevent tumors. And as soon as you stop the ginseng, the cancer cells grow again. It's not a permanent treatment, it's not a cure. But what was interesting is when you give ginseng together with traditional chemotherapy drugs, you significantly increase the efficacy of the chemotherapy drugs. That's where I think there may be some therapeutic potential—not with ginseng itself, but with ginseng used in combination with traditional drugs. That's what we're seeing—some really really neat stuff!"

A problem that confronts Laura Murphy, and all other ginseng researchers, is that the drug they're studying is not a single substance, but an herb—and an extraordinarily complicated one at that. Most medicinal plants contain a single active ingredient. Ginseng contains at least twenty—there's disagreement about the exact number.

The first person who attempted to isolate the active component in ginseng was a chemist named S. S. Garriques, who discovered in 1854 that the roots contain a group of chemicals called saponins, the name derived from a term meaning "soapy." The saponins were later confirmed as the active ingredients in the plant, and named ginsenosides. These substances make up about 3–6 percent of the weight of the dried root, and cause the foaming that occurs when ginseng tea is brewed from roots.

During the 1960s, two groups of Japanese researchers isolated thirteen different ginsenosides, designated Ra, Rb, Rc, and so on, with the second letter indicating the sequence in which they separate out in a process called thin layer chromatography. Asian and American ginseng contain slightly different collections of ginsenosides, in slightly different proportions.

A key question, which remains unanswered, is: Exactly how does ginseng work on the body? Are its medicinal effects the result of individual chemicals in the plant? Or do they depend on the synergistic interaction within the root as a whole?

Chinese medicine has always postulated that ginseng's effects are the result of its stimulation of the *chi* energy in the body. By increasing the flow of *chi,* at the same time it builds up anemic blood, increases the appetite, improves respiration, and strengthens the body's natural resistance to disease.

A Western version of this concept is that ginseng has an effect on the body's regulatory mechanisms—an idea that goes back to the Soviet "adaptogen" theory, which says that ginseng's main effect is to help the body adapt in response to stress. British pharmacologist Stephen Fulder, in his *Book of Ginseng,* outlines a more rigorous theory of how this could work.

Fulder argues that ginseng affects the body by regulating the functioning of hormones. Hormones are the body's chemical "messengers," carrying signals that control various body processes. For example, growth hormone controls the growth and development of tissues, while the hormones estrogen and testosterone regulate reproductive and sexual functions. A hormone that is familiar to non-scientists is adrenaline, which regulates the quick response to stress—the body's "fight or flight mechanism"—producing familiar effects such as sweating and a faster heart rate. These hormones are produced by the adrenal glands, which are controlled by a part of the brain called the hypothalamus.

In experiments with laboratory animals, Fulder found that ginseng has a strong effect on the adrenal glands. Animals who were fed ginseng had a faster adrenal response when subjected to stress, and then returned to a normal state more quickly. He further found that ginseng improved the ability of the hypothalamus to react to stress feedback from the body.

This may not sound earth-shaking, but many of the most common killers of the modern world, including heart attacks, stroke, and cancer, have been related to stress. Stress hormones can inhibit the functioning of the immune system. As Fulder writes, "Many, many medicines are known to change bits of the body machinery, but there are few indeed that can increase the smooth running of the whole."

Laura Murphy, on the other hand, is looking at the microscopic level. She is now trying to pinpoint the exact mechanism through which ginsenosides act on cells, and has just received a prestigious grant from the National Institutes of Health to look into this. "We're going to do some sophisticated molecular biology techniques to try to dig into how ginseng is acting at the level of the genes to alter cancer cell proliferation. We're going to get pretty heavy-duty into the mechanism. Once we get into the mechanism, I think then we can use the ginsenosides for medicinal purposes."

Meanwhile, in a very different kind of lab, Father Jack Kieffer is working on a very different kind of ginseng research. For starters, the little green shed where he works is probably one of a very few scientific laboratories in the world that's heated by a cast-iron woodstove and which looks down a forested hillside to a tumbling mountain river. His lab bench is made from nailed-together two by fours. When his day's work is finished, he hangs his impeccable white lab coat on a nail in the wall and walks down a stone path to the sound of birdsong. Though it sounds implausible, some of the ginseng extracts used in Laura Murphy's breast cancer research were produced right here.

Kieffer is a thin, older man with a fringe of gray hair, whose speech holds long pauses for deliberation. His tiny lab, no more than eight feet by twelve, stands on the peaceful grounds of ASPI's Rockcastle River Demonstration Center, where Syl Yunker taught his workshop for aspiring ginseng growers. In this unlikely location, Kieffer is now trying to isolate the individual ginsenosides contained in the leaves and stems of the ginseng plant—an idea which has been almost entirely overlooked by other researchers, who have focused almost exclusively on the roots. More than just intellectual curiosity drives his work. In raising ginseng roots as a cash crop, growers must wait a minimum of five years to get any income at all. But Kieffer has discovered that the plant's leaves also contain many of the same active substance as the roots, and ginseng leaves can be harvested every autumn without harming the plant. If a medical use for the leaves can be found, this could provide a source of income for growers to sustain them during the long years of waiting to harvest the first roots.

Kieffer has been working for several years now on this project, which originally grew out of his work to protect the Appalachian forest from clear-cutting and strip-mining. The capacity to grow a high-value crop such as ginseng makes intact forest much more economically valuable than any one-time harvest of old-growth timber could ever be. When the Appalachian Ginseng Foundation was established to support small ginseng growers, Kieffer took over a tool shed to use as his laboratory. Necessary instruments and glassware were donated, scrounged from Appalachia-Science in the Public Interest, or bought secondhand at a surplus auction at the University of Kentucky, fifty miles up the road.

The procedure that Kieffer describes to me is an intriguing mix of homey and high-tech. First, the ginseng leaves are dried in a solar food dryer (otherwise used for veggies from the center's organic garden) and then broken up using a kitchen blender. The powder is combined with water and alcohol, and mixed with a magnetic stirrer at room temperature (which, I later read in an ASPI report, "has varied from 30 to 60 F").

"It's like how we digest our food," Kieffer tells me. "We have to chew it up to break the cell walls open so we can get the nutrients out."

Next, the mixture goes into a centrifuge, which spins it to separate out a solution of the ginseng chemicals. After that liquid is concentrated further by drying, it goes through a process called thin layer chromatography, in which drops of the solution are dried on a glass plate coated with silica gel in order to divide them into their components.

"If you drop beet juice on a towel," Kieffer explains, "the stain will spread out in different colors. That's the principle behind this." The solution is dropped on the plate and dried, and then the plate goes into a tank of butanol, ethyl acetate, and water. The liquid mixture carries the chemicals up the plate, with different ginsenosides moving at different rates, spreading out into different spots. Exposing the plate to iodine vapor makes the spots visible. Kieffer then circles the spots gently with a pencil. Each spot can then be removed and dissolved again, in water or alcohol, producing a solution containing a single ginsenoside.

By using this process, Kieffer is able to separate seven or more ginsenosides from a batch of leaves. It seems awkward and slow, but he assures me that it works well—he manages to extract more ginsenosides from a sample than Dr. Murphy gets in her state-of-the-art university lab. Given enough time and equipment, a tiny lab like his could produce salable quantities of individual ginsenosides—now priced by commercial laboratories at around $17,000 per *gram.* That's all hypothetical, though. "I've never been in the money dimension of the world," Kieffer laughs. "That doesn't interest me."

What does interest him is a surprising discovery that Murphy has made: ginseng leaves are even more effective than ginseng root in stopping the growth of breast cancer. She found that the leaf extract is ten times as potent as the root extract, and appears to actually kill the cancer cells rather than just stop their growth—"An encouraging piece of data," he calls it. Furthermore, it seems that the concentration of ginsenosides in the leaves increases as the growing season goes on, meaning that leaves can be left on the plant to fuel the root's growth until late in the fall.

This has sparked some interesting "alternative" ideas around the Appalachian Ginseng Foundation: solar dryers for ginseng leaves, growers marketing packaged ginseng-leaf tea bags directly to consumers via the Internet—anything to help the little guy get through to the first harvest.

Ginseng is an extremely tricky plant for scientists to work with, though. Though the twenty (or so) ginsenosides are all considered to be the medicinal substances, each ginsenoside has different effects—in fact, some cancel each other out. For example, one ginsenoside lowers blood pressure, while another elevates it.

"It's like an herb with tons of drugs in it," Laura Murphy tells me over one last coffee refill at Denny's. She has found evidence that some single ginsenosides are even more powerful against cancer than ginseng given as a whole.

"We showed that some of the ginsenosides we've isolated—one of them in particular, ginsenoside Rc, by itself was more effective than the whole root. That would suggest that there must be other components in the ginseng plant that are actually interfering with the action of that compound." What scientists observe when testing the ginseng root or the leaves, she says, is a net effect of all the different components. If they can tease apart which components have just an anti-cancer effect, and then combine those, they may have a very potent mixture. Preliminary studies have looked at individual ginsenosides, and now combinations of ginsenosides are being investigated. Studies found that mixing ginsenoside Rc, which has anti-cancer effects, with ginsenoside Rb1, which has the highest concentration in the root, but no effect on the cancer cells, produced a synergistic additive effect.

"So what we're trying to do now," Murphy explains, "is find out what combination of ginsenosides you can put together that will really have the maximum inhibitory effect. It's tricky, because you've got twenty different drugs in the root. Finding out what drugs are interfering with the really potent ones is what we want to do."

"Well," I ask, "supposing you find some substance that is really powerful against cancer—what's to prevent some big pharmaceutical company from patenting it, and synthesizing it? Or can't that be done?"

"Yeah, that can be done." She sounds totally unconcerned.

"Don't you worry about that?"

"No, not really," she laughs. "That's going to *help* people. And I hope that's where our research actually goes. I hope that it does interest the pharmaceutical companies into looking at potential therapeutic uses for ginseng, and taking apart ginseng and looking at what parts have the anti-cancer effects and synthesizing them. I had a chemist in my laboratory for two years who was actually helping me to synthesize ginsenosides. So we're totally able to do it, and it's cheaper than actually purifying the ginsenosides from the ginseng plant. We tested them, and they were very potent."

But wouldn't this cut all the farmers out of the picture?

She assures me that will never happen. "Yes, that *is* what Western medicine would prefer, to have a single drug, and be treating a disease with a single drug. But ginseng is going to Asia, it's the Asians who consider ginseng in toto. It's all the ginsenosides and whatever else is in the plant, be it physical or spiritual, that's responsible for its restorative and medicinal effects. You've got two very different cultures here."

The main market for ginseng, she believes, will always be in Asia, where the demand will always be for the whole root. And if Western research proves that ginseng actually does prevent cancer, it will send throngs of consumers to the drugstores in search of over-the-counter ginseng products. Even if pharmaceutical companies get on board and start synthesizing ginseng components as treatments for specific diseases, this is likely to raise demand. "It's still going to increase ginseng's marketability, because you and I are going to be taking it because it has those constituents."

Still, for Laura Murphy and other researchers, it's an uphill battle getting ginseng recognized as serious medicine. "There's just this bias against the concept of herbs as an alternative, or as a complementary medicine. And I'm rising to the challenge. I have a goal to educate the scientists and see if we can get this more accepted. Once we get the papers published, once we get this out there, there'll be more money available for research, and when the money is there for research, we're going to get evidence-based facts on herbal remedies and ginseng in particular. And once that evidence base is out there, that's when the public is going to say OK, scientists and doctors are saying I can use this. It's got medicinal effects."

"How long will that take?" I ask her as I close my notebook and pick up the check.

"Boy, I don't know," she says. "Too long. But it's getting better."

This doesn't mean that practitioners are waiting for definitive research results to come in before they start giving ginseng to their patients. Along with several millennia of traditional Chinese physicians, Western health practitioners also treasure the herb.

Andrew Bentley, an herbalist with a clinical practice in central Kentucky, uses ginseng in combination with other herbs to treat a variety of problems. A thunderstorm crackles the phone line as he lists them for me in his thoughtful, deep voice: "It's sometimes helpful for people who feel depleted or don't have a lot energy, or people who don't digest their food well. It can help with infertility, more often with men than in women, sometimes with both depending on what the cause is. Sometimes it's helpful as an adjunct along with other things for people who have cancer."

What about specific people who have been helped by ginseng?

He thinks a moment. "One was a woman twenty-nine years old, and she was experiencing extreme daytime sleepiness, feeling like she was going to fall asleep while she was walking around. I had her stop using caffeine, which she had been pretty reliant on, and take ginseng six days a week. Right away she stopped having the extreme sleepiness, and in about two weeks she had a pretty normal energy level. She took it six days a week for a period of two months. That was a couple years ago, and she's still doing fine, not having that kind of symptoms any more. That's a real simple example."

Because he normally uses ginseng in conjunction with other herbs, it's harder for him to think of a case where the outcome is clearly due to ginseng. "I did have one couple I saw, where the man had irregular sperm morphology, his sperm weren't shaped right. I had him take ginseng for three months, and then stop taking it for a month, and then take it for three more months. And they conceived." It's hard to think of a more dramatic result.

An herbalist can give ginseng as a fluid extract made by soaking the roots in alcohol, or as a syrup made the old-fashioned way: "You take the roots and you put them in water and you boil it, and then you let them sit for three days. Then you take the roots out and you boil it down to about half the volume that it was, and then you take half of that volume of honey and mix it in there, and that keeps it from spoiling. And then you put it in sealed containers." And, of course, there are other commercial preparations available: tablets, capsules, slices of candied root.

For Bentley, one of ginseng's most important assets is its venerable history. "We have documented clinical history of use going back for thousands of years on ginseng," he points out. "That's really something that people consider to be the most reliable source of information on a therapeutic substance, how much history of use you have, how long it has been around. We know better what to expect from a drug like aspirin than from a newer one. It's the same with ginseng. It's been around so long that we have a really good idea of what it does and doesn't do."

Ginseng research is going on in many forms, on many fronts. Up in Wausau, the Ginseng Board of Wisconsin has funded a study to investigate the potential use of ginseng to help diabetics control their blood sugar. In Canada, researchers have found that ginseng can help to lower people's cholesterol levels. In the last two decades, other studies have looked at ginseng's immunological effects on people with HIV, at its use in treating severe chronic respiratory diseases, its effects on fatigue, depression, menopausal symptoms, erectile dysfunction, and just about every other problem that plagues the human organism. Most of these were small scale studies, requiring further testing, and some had negative or inconclusive results. But the groundwork has been done. A German government commission in the 1990s, after intensive investigation of its properties and a thorough review of the research, included it on the official list of approved herbal medications. Ginseng is currently listed in the national pharmacopeias of Austria, China, France, Germany, Japan, Switzerland, and Russia.

Consider this: the genus name of ginseng, *Panax,* is derived from the Greek *pan* (all) *akos* (cure), the source of the word panacea. The plant's scientific name means, literally, cure-all.

CONCLUSION

GINSENG DREAMS

In Siberia, the last of the wild ginseng hunters lived on into the twentieth century. Long after the rest of the world believed that wild Asian ginseng had vanished, they were still gripped by a kind of fever, and wandered the taiga of Northeast Asia in a quest for the roots that remained.

In a series of three obscure books published in the 1930s, a Russian explorer and writer named Nikolai Baikov recorded the wretched lives of these Chinese hunters who roamed the forests from infancy to old age, their eyes fixed on the ground, searching for a rare plant. His observations were translated and preserved in an only slightly less obscure book from 1975, Andrew Kimmens's fascinating *Tales of the Ginseng.*

"They are recognized by their dress and equipment," Baikov wrote, "an oiled apron which protects them from the dew, a long stick with which to part the leaves and grasses, a wooden bracelet on the left arm, and a badger skin hanging from behind their belt which lets them sit on wet ground. They generally wear a conical hat of birch bark, tied under the chin by a thong, and shoes of tarred pigskin."

Every summer, they set off on their solitary way, walking hundreds of miles through trackless wilderness. To mark their

---

Previous page: Grower Ginger Shelby shows off some of her top-quality ginseng roots, all raised organically and packed in natural materials.

passage, they would leave signs like broken branches or cuts in tree bark to inform their brother hunters that an area had already been explored. Upon finding a root, they would say a prayer, reminding the spirits that they had come with a pure heart and asking for their aid. Then, before digging, they would carefully study the plant and its particular situation. Using digging tools consisting of a set of spades made from bone, and a long, flexible knife, they would work gently to avoid harming the limbs and fingers and toes of the root growing deep into the earth. When the root came free, it was wrapped in thin paper, then laid in a birch bark basket.

Theirs was a miserable existence. In the extreme weather of Siberia, the hunters had no shelter other than caves, rocks, or the branches of a cedar tree. Bandits lay in wait for them as they neared the towns and settlements. Even if they managed to bring back a sizable number of roots, they were at the mercy of the buyers and brokers and trading houses, and none of them ever grew rich enough to leave that life.

But they didn't give up hope. The hunters built shrines to the spirits of the mountains, renounced the world, and gave themselves over to the gods' protection. This earned them the respect of the settled tribes of the taiga, some of whom began to cultivate ginseng themselves. All viewed the plant as sacred, and no one would dare to touch another's roots. Sooner or later, anyone who dared to steal ginseng would be found dead in his cabin, no matter how far away he lived. Whether this was due to comradely revenge, or the magical powers of ginseng, Baikov didn't explain, but righteousness prevailed in the end.

For these hunters, Baikov wrote, ginseng meant far more than money. "The plant is a kind of stimulant of justice and good, a symbol of the balance between natural, creative forces. The Chinese idea is that the root itself contains a part of the Great Spirit: ginseng is the seed of light, the source of movement, invisible life, and universal energy, inspired, possessing the supernatural force which distinguishes divinity."

Where once there had been over eighty thousand ginseng

hunters roaming the wilds of Manchuria, by the 1930s, when Baikov met them, only ten thousand remained. No one knows when, or if, the last one gave up. It would not surprise me to hear that out in the roadless vastness of Siberia people are still chasing the ancient root.

Does ginseng have a future in America? This is my final question for all the people who share their stories with me. The answer is nearly always: Yes—maybe.

Scott Persons, author of the most widely used manual for small-scale growers, hedges when I ask him that. But obviously, if he's made the effort to produce an entirely new book, he thinks it will find an audience.

"I see conflicting trends," he tells me. "I'm still planting it, but it's not that I think things are going to get a lot better, and it's not that I'm real worried that things are going to get a lot worse. As the younger generation comes along in the Orient, there are fewer people that are born and raised by parents who think that ginseng is essential. And younger people do not have the same reverence for it that their parents and grandparents did. The market in the Orient for ginseng is still good, but it's not blazing. So that is a potential problem, that their population is becoming more modern. They want to see the research that this specific chemical has that specific effect, rather than you take the whole root and there's this holistic balancing that occurs, which raises your ability to cope with most any stress on the body."

He feels that the future market may be slanted more heavily toward the West, toward cultures that have historically viewed ginseng as superstition, a mere curiosity at best. "There's a growing appreciation for herbal medicine throughout the world. Partly there's an economic incentive to look at these things, and partly there just seems to be a more open mind. Twenty years ago, when I first got into ginseng, the culture that I lived in thought it was a joke, actually. There's more open-mindedness about it, a sense that 'there just might be something here.' We thought we were so smart, but maybe we weren't. Maybe, over thousands of years, a billion Chinese were pretty smart, too."

Probably no one in the ginseng world has a broader and deeper view into the future Chinese market than Paul Hsu, the biggest grower of American ginseng in Wisconsin, and the United States, and very possibly the world. His companies account for 20 percent of the U.S. harvest. He spends many weeks every year traveling in Asia, visiting his joint ventures in China, his offices in Hong Kong and Malaysia. It takes me more than two months just to schedule a phone conversation with him.

Hsu was born and raised in Taiwan, and came to the United States more than thirty years ago to study social work at the University of Denver. When I finally get to ask him how he got started in ginseng, I'm expecting a tale of many generations in a venerable family business. Not even close.

"I used to work for the State of Wisconsin," he tells me, his quick words still touched with a light Chinese accent. "In the Division of Health and Social Services, mainly in foster care and child welfare. One day a colleague of mine said, 'Oh, I read an article about American ginseng, do you know anything about it?' I said, no, I'd heard about it, but my mother used to use Korean ginseng." So Hsu read the article, which had appeared in, of all places, the *Mother Earth News*—the "back to the land" journal that launched a thousand hippie homesteads—and found himself captivated.

There are echoes of the Korean legend of Kang and the Mountain Spirit. At that time, Paul Hsu was very concerned about his mother's health—after bearing and raising fourteen children, she was weakened by ulcers, diabetes, and arthritis. The *Mother Earth* article convinced him to investigate American ginseng and send her some. Marathon County bordered the region that he covered for the state, inspecting foster homes, and one afternoon he took a couple of hours off and visited some farms, buying two pounds of roots to send to Taiwan. He hoped for the best, figured he had done all he could.

What came next took him by surprise. "My father wrote me a letter six months later saying, 'You won't believe this, but your mother's health has improved a lot. She used to look pale, but her color is coming back now. Her arthritis is better, and her diabetic condition is almost cured, and indigestion is no problem—she used to eat one bowl of rice, right now she eats three bowls of rice every meal.'" The next year, he went home to celebrate his parents' fiftieth wedding anniversary, and found his mother's health vastly improved. "It made a believer out of me," he says.

Hsu's wife had just finished her nursing degree, and with her income to fall back on, they decided to take a huge gamble. He quit his state job and started a ginseng business. His timing was impeccable. "I started as a marketer," he tells me. "I did mail-order first. In 1974, the toll-free phone number was being popularized then, the credit card was popularized, all that was playing into what I was planning. I started with mail order because the market at that time was for the Chinese population in the United States, especially in the secondary and tertiary cities. The major cities have Chinatowns where ginseng is available, so my target at that time was cities like Dallas, Miami, Denver. . . ." His business took off like a shot. "Then I started exporting," he recounts. "I targeted the Taiwan market, so in five years I had 85 percent of the market share in Taiwan. Then I would go into Hong Kong, Southeast Asian countries, Singapore, Malaysia, then China." Hsu Ginseng now has eleven offices in the United States and Asia, with 350 employees, and does $30 million of business every year.

And as for the Chinese, he tells me, their enthusiasm for American ginseng is only growing. "Right now, people are getting more affluent, they are getting more concerned about their health, and there's some kind of attraction because it's imported from the United States. There's a prestige. Things that come from the U.S. have a different perceived value. Now it's not only just a treatment, a regimen for disease, it's used for health maintenance, as a health tonic now. Especially American ginseng is used for tonic. I would say 10–20 percent is used as a treatment, and 80–90 percent as a health maintenance thing, similar to a vitamin or a food supplement. Chinese ginseng, on the contrary, probably 80 or 90 percent of it is used as a prescribed herbal medicine."

Hsu himself consumes plenty of ginseng, and is convinced of its powers. He tells me he's had exactly two colds in thirty years, that at age sixty-three he can still run two miles a day—or could, if he weren't "lazy."

He also feels that population growth and increasing affluence in China will boost the market for American ginseng. In the past, he estimates, only the 1–5 percent of Chinese at the top of the income scale could buy ginseng; now 15–20 percent of it can afford it, putting it within the reach of the growing upper-middle class.

I'm stunned to hear, after all this, that Paul Hsu, one of the biggest and shrewdest ginseng growers in history, is in fact slowly getting out of the business. The reason? Competition from China. In recent years, the Chinese have begun cultivating large areas of American ginseng, grown from American seeds planted in China. Given the right conditions, the plant grows just as well, though Hsu argues that the quality of the roots is far lower. And, needless to say, with China's cheap labor, the cost of production is also far lower. Large-scale American farmers like the guys in Marathon County are finding it more and more difficult to compete. They try to highlight the quality of Wisconsin roots, but to many Chinese who have only recently been able to afford ginseng, price is the only thing that counts.

"In Chinese herbology," Hsu explains, "the locale specificity is very, very important. Just like grapes produced in Champagne, France—that grape transported to Napa Valley in California may look like the same grape, but when you make champagne it tastes different, and therefore the price is different And the problem is the consumer doesn't understand that totally. They say, 'Hey, we've got champagne for $3.99, instead of $29.99 from France.'"

Ginseng prices are declining all over the world as production goes up. Twenty years ago, I learn, ginseng was four times more expensive than it is now. In real terms, it currently fetches only one eighth of the price it used to. The business is not only more competitive, it's less lucrative. "It used to be a good crop, and right now it's very tough. I've been in this business thirty years. It's very different."

For Hsu, the solution is to depend less on ginseng. "If I weren't diversifying, I wouldn't be here today. I haven't grown any larger since 1992. So it will be twelve years already I've concentrated on marketing, that's why I'm still here. After 1992 I got into many more products—herbal products, Chinese health products we developed, plus high-priced seafood items. So, we are pretty diversified. Right now ginseng is only 60 percent of our business. In five years, it will probably be half."

And yet, a minute later, he tells me that his son has just started studying at Harvard for his MBA. "And if he shows some interest . . . ," he says. His voice lifts with hope.

Whatever powerful force it contains—ginsenosides, crystallized lightning, yin essence—American ginseng seems to ignite dreams in everyone who touches it, in everyone it touches.

It will unite the heavens and the earth. It will preserve the Emperor in virtuous longevity. It will win the souls of the Iroquois for Jesus. It will earn me the money to buy a fleet of schooners. It will save my dying mother. It will buy me a rich homestead in that newfound land on the far side of the Blue Ridge. It will be the driving force of a new and lucrative industry. It will make me filthy, stinking rich. It will relieve the grippe, nervous debility, and weakness of the stomach. It will bring me and my brothers enough money to buy a breeding pair of foxes. It will strengthen the flow of *chi.* It will buy piles of Christmas presents for all the grandbabies. It will make this poor Wisconsin county famous half a world away. It will pay my tax bill, my mortgage. It will grow forever if we just pass the right laws. It will end poverty in Appalachia. It will take me back to great-grandpa's honest way of life. It will save the forests, stop the strip-mining, end the clear-cutting. It will lower my blood pressure, control my diabetes, spark my energy. It will cure cancer. It will make us all immortal.

I end with a coda, a truth-is-stranger-than-fiction scene that no decent novelist would ever try to get away with.

Just over a month after I visit Dr. Laura Murphy, and stare in horror and fascination at the lovely, deadly sequin-cells under her microscope, I'm diagnosed, myself, with a rare form of cancer. My only symptom is a pressing pain in my gut that suddenly quits after three days, never to return. I go through stomach x-rays, an ultrasound, a nuclear scan. A blood test finds my liver enzymes elevated, a sign that its cells are dying. I have CT scans with dye shooting hot through my veins, then an MRI, then an ERCP, a nasty procedure involving a camera passed down my throat to just north of my belly button. Everything comes up inconclusive, until the CT-guided needle biopsy.

There's a malignant tumor growing inside my bile duct, a tiny but vital piece of plumbing inside the liver. It's a cancer almost unknown in North America, though common in East Asia, and no one can figure out why I got it.

I'm lucky, extremely lucky. After surgery to remove my bile duct, half of my liver, and my gall bladder, I'm given a clean bill of health, with no chemotherapy or radiation treatment needed. But there's a caveat. This is not a cancer with a good survival rate. If it comes back, says my oncologist (a phrase I never expected to use in this lifetime), it will be in my liver.

A few weeks post-op, I'm feeling strong enough to sit up at my computer, and I send an email to Murphy, telling her what's happened to me. She writes back, encouraging: "There is human evidence that ginseng users exhibit a lower incidence of gastrointestinal cancers, including liver cancer. In cell studies, ginseng and ginsenosides have been shown to inhibit the proliferation of human liver cancer cells."

I've touched it, and now I have my own ginseng dream. It will keep me going.

BIBLIOGRAPHICAL NOTES

Much has been written about ginseng in general, and American ginseng in particular, but most of this information is scattered in obscure publications or hidden in places that few would think to look. Ginseng has had a number of "heydays"—during the earliest boom of the ginseng trade in the 1770s; in the era when it was first successfully cultivated, in the 1910s; throughout the back-to-the-land days of the 1970s; and (arguably) now, as the rising standard of living in China brings ginseng consumption within reach of an ever-growing segment of the country's billion-plus population. Each of these peaks of interest triggered a corresponding boom in publications on the topic—with earlier efforts largely forgotten.

But an interesting difficulty in carrying out research on ginseng is that works on this topic must rank among the most frequently stolen items in the nation's libraries. On a number of occasions, I closed in on some promising-sounding old book or intriguing pamphlet only to find the folder in the vertical file empty, or an unauthorized gap in the call numbers on the shelf.

EARLY HISTORY OF AMERICAN GINSENG

The life and fortunes of Penn Kirk and his *Ginseng Journal and Goldenseal Bulletin* are drawn directly from the volumes of that journal preserved at the Lloyd Library and Museum in Cincinnati. That institution, devoted entirely to herbal medicine and its history, is a delightful treasure-house for anyone interested in the field.

Penn Kirk was just one of the thousands of ginseng cultivators who flourished and then faded away in early twentieth-century North America. Few of them left any record of their work, though. Two detailed (if meandering) reminiscences by early ginseng growers and promoters are Val Hardacre's *Woodland Nuggets of Gold* (New York: Vantage Press, 1968), which records every detail of his life down to the names of his wife's prize-winning show cats, and A.R. Harding's more businesslike *Ginseng and Other Medicinal Plants* (1908; reprint, Columbus, Ohio: self-published, 1972).

For a broader outlook on the history of ginseng, a useful overview of the development of the North American ginseng trade with China can be found in Alvar W. Carlson's paper "Ginseng: America's Botanical Drug Connection to the Orient" (*Economic Botany* 40, no. 2 [1986]: 233–49). Details of the discovery of American ginseng in Quebec are given in "Ginseng: Jesuit Connections, Appalachian Hope," by Al Fritsch (*National Jesuit News,* Oct. 1998, pp. 3–4), and "Ginseng and the Royal Society Connection," by David Bellamy and Andrea Pfister (in *World Medicine: Plants, Patients, and People* [Oxford: Blackwell, 1992]). *Tales of the Ginseng,* by Andrew C. Kimmens (New York: Morrow, 1975), is a fascinating anthology of rare material about ginseng and its folklore in both Asia and North America, with a copious annotated bibliography containing lengthy excerpts from still more sources.

WILD GINSENG

A detailed discussion of the habitat, characteristics, and status of wild American ginseng is provided in a paper by Roger C. Anderson et al. titled "Wild American Ginseng" (*Native Plants Journal* 3, no. 2 [2002]: 93–97). Wonderful, atmospheric audio excerpts from Mary Hufford's interviews with diggers of wild ginseng in West Virginia can be heard on the Library of Congress's searchable American Memory website (http://memory.loc.gov/ammem/). Hufford describes her work on forest health and discusses the significance of ginseng in her paper "American Ginseng and the Idea of the Commons" (http://memory.loc.gov/ammem/cmnshtml/essay1/), which also includes photos and interview excerpts.

CULTIVATED GINSENG

If you ask ten ginseng growers about the best method to use, you will hear at least twelve conflicting (and vehement) opinions. There is no one authoritative source, and trying to get to the bottom of the endless debates means wading through voluminous and contradictory information. That said, a comprehensive reference manual for the modern-day grower, whether in fields or in forest, is W. Scott Persons's *Growing and Marketing Ginseng, Goldenseal, and Other Woodland Medicinals* (Fairview, N.C.: Bright Mountain Books, 2005), written with Jeanine Davis. A slightly different approach is outlined in Kim Derek Pritts's *Ginseng: How to Find, Grow, and Use America's Forest Gold* (Mechanicsburg, Pa.: Stackpole, 1995).

Plenty of other practical information is in print. The Ginseng Board of Wisconsin (http://www.ginsengboard.com/Index.htm) issues periodic fact sheets and reports geared to large-scale cultivators. *Ginseng: A Concise Handbook,* by James A. Duke (Algonac, Mich.: Reference Publications, 1989), provides a somewhat more skeptical perspective on the merits of ginseng as medicine, while still offering plenty of hard facts on ginseng as a crop. The USDA has produced information for growers including "Growing Ginseng" (Farmers' Bulletin Number 2201, 1978).

The saga of the Fromm Brothers and their Marathon County ginseng empire is related in near-heroic terms by Kathrene Pinkerton's *Bright with Silver* (New York: William Sloane, 1953). Descriptions of ginseng cultivation in the nineteenth-century Korean Peninsula can be found in Isabella Bird Bishop's fascinating 1897 memoir *Korea and Her Neighbors,* reprinted by Yonsei Univ. Press of Seoul (1970).

WOODS-GROWN GINSENG

W. Scott Persons's manual, listed above, contains a detailed discussion of various approaches to growing ginseng in the forest. The Appalachian Ginseng Foundation and Appalachia-Science in the Public Interest have issued a number of publications promoting and describing Syl Yunker's virtually wild method, including "Ginseng in Appalachia" (ASPI Technical Series TP38, 1996). These can be downloaded at their website, http://www.a-spi.org/pub.htm. Various state government agencies in the Appalachian region have also produced practical guides for those hoping to grow ginseng in the forests of their states, such as the West Virginia University Extension Service's "Woods-Grown Ginseng" (1995) and the North Carolina Cooperative Extension Service's "Care and Planting of Ginseng Seed and Roots" (Horticultural Information Leaflet 127, 2000).

GINSENG POACHING

The *Washington Post* reported on the Shenandoah National Park operation and the controversy surrounding it on June 1, 2004, in a front-page story titled "Virginia Sting Targets Trade in Bear Parts, Ginseng." But despite the extent and impact of the problem, there has been little ongoing press coverage of ginseng theft. My own article on the subject in *Mother Jones* magazine, "Root Rustlers" (July/August 2002, p. 18), was one of the few that have appeared in a national magazine. For a detailed report of the struggle against poaching in the Great Smoky Mountains National Park, see Burkhard Bilger's article in the *New Yorker,* "Wild Sang" (July 15, 2002, pp. 38–45). It includes an account of Jim Corbin's tireless efforts to develop a ginseng marking system—though Corbin was rather distressed that the author described him as padded in "baby fat." ("I still run five miles every day," he complained to me, "so surely to goodness I'm not cushioned with TOO much baby fat. He was a little old skinny fellow. . . .") Possible deterrents to poaching for small growers are discussed in the Appalachian Ginseng Foundation's leaflet "Ginseng Crop Protection" (2001).

MARKETING AND THE STATUS OF THE SPECIES

Ginseng dealers are generally a pretty close-mouthed group, and none of them that I am aware of have ever set their point of view down in print. The U.S. Fish and Wildlife Service's current annual finding on the status of American ginseng can be downloaded from http://international.fws.gov/animals/ginindx.html. Arguments for an end to the wild ginseng trade are presented in "Moratorium on Wild American Ginseng Exports" by the Appalachian Ginseng Foundation (ASPI Technical Series TP57, 2000).

MEDICINAL USE OF GINSENG

Copious literature has been produced about the role and effect of different types of ginseng within both the Chinese and Western medical traditions. *Facts about Ginseng: The Elixir of Life* by Florence C. Lee (Elizabeth, N.J.: Hollym, 1992), from the Asian perspective, is more science-based and less sensational than its title would suggest, but deals mainly with Asian ginseng. Herbalist Paul Bergner discusses the use of American ginseng within the context of Traditional Chinese Medicine in *The Healing Power of Ginseng and the Tonic Herbs* (Rocklin, Calif.: Prima, 1996). The works of British pharmacologist Stephen Fulder (including *The Ginseng Book: Nature's Ancient Healer,* Garden City Park, N.Y.: Avery, 1996), provide lengthy summaries of Soviet-era research, and present his own idiosyncratic theories about what ginseng actually does. A plethora of research studies around the world have investigated ginseng's effectiveness in treating or preventing everything from radiation sickness to hangovers, but an evaluation of the validity of these studies is far beyond my capabilities, and the scope of this book. For anyone who wants to dive in, one place to start is an online medical database such as Medline, where a recent search on the keyword "panax" (the scientific name of ginseng) turned up 1,949 different papers. Laura Murphy's initial work on ginseng and libido was published in "Effect of American Ginseng (*Panax quinquefolium*) on Male Copulatory Behavior in the Rat" (*Physiology and Behavior* 64, no. 4 [1998]: 445–50).

By the time you hold this book in your hands, there will doubtless be dozens more breakthroughs in ginseng research—and just as many opposing experts lined up to debunk them. But whatever the results, I have no doubt that hundreds of millions of Chinese will still be placing their trust in ginseng's life-giving powers, with increasing numbers of Westerners joining them.

INDEX